旋转、逆转火焰的热特性和污染排放特性

甄海生 著

·北京·

内 容 提 要

本书旨在研究以液化石油气（LPG）为燃料的一种基于旋流的、超稳定燃烧的逆转扩散火焰（IDF）的热、污染和传热特性，主要以实验性质为主。第1章为概况和动机；第2章是研究的背景，其中包含与逆扩散火焰相关的前期工作的详细文献回顾；第3章介绍了实验装置、仪器和测量方法；第4章到第7章，分别介绍火焰外观、火焰结构、火焰温度场、火焰稳定气体种类、火焰冲击热传递、火焰比较等；第8章是旋转火焰的数值模拟，计算结果作为实验研究的补充；最后，在第9章中对本书的研究结果和结论进行了全面的讨论，并对今后的工作提出了建议。

本书是研究旋转、逆转火焰的学术类专著，具有较高的学术价值，对于专门从事研究旋转、逆转火焰的科研人员及学术类研究群体具有非凡意义。

图书在版编目（CIP）数据

旋转、逆转火焰的热特性和污染排放特性 / 甄海生著. -- 北京 : 中国水利水电出版社, 2018.4 （2025.4重印）
ISBN 978-7-5170-6377-3

Ⅰ. ①旋… Ⅱ. ①甄… Ⅲ. ①火焰－燃烧化学－研究②燃料工业－污染防治－研究 Ⅳ. ①O643.2②X784

中国版本图书馆CIP数据核字(2018)第064207号

责任编辑：陈　洁　　　　**封面设计：王　伟**

书　　名	旋转、逆转火焰的热特性和污染排放特性 XUANZHUAN NIZHUAN HUOYAN DE RETEXING HE WURAN PAIFANG TEXING
作　　者	甄海生　著
出版发行	中国水利水电出版社 （北京市海淀区玉渊潭南路1号D座　100038） 网址：www.waterpub.com.cn E-mail：mchannel@263.net（万水） sales@waterpub.com.cn 电话：（010）68367658（营销中心）、82562819（万水）
经　　售	全国各地新华书店和相关出版物销售网点
排　　版	北京万水电子信息有限公司
印　　刷	三河市同力彩印有限公司
规　　格	170mm×240mm　16开本　12印张　206千字
版　　次	2018年4月第1版　2025年4月第3次印刷
印　　数	0001—2000册
定　　价	48.00元

作者简介

甄海生，男，1979年出生于河北省，吉林大学学士，哈尔滨工业大学硕士，香港理工大学博士。2006年赴香港从师Leung Chun Wah、Cheung Chun Shun教授，研究火焰燃烧热特征，污染物排放及冲击换热10余年。研究对象和内容涵盖预混、非预混，层流、湍流，旋流、非旋流，富氧、贫氧和富氢火焰；传统石化燃料和可再生新燃料；火焰燃烧的热释放、污染物产成机理和控制，洁净燃烧和节能减排技术；火焰热释放和热能高效利用；以及燃烧噪音机理和防治。

前　言

本书旨在研究以液化石油气（LPG）为燃料的一种基于旋流的、超稳定燃烧的逆转扩散火焰（IDF）的热、污染和传热特性。开发的旋流燃烧器可产生高度旋转 IDF，其几何涡旋数高达 9.12，且火焰内部形成回流。旋转 IDF 的控制参数有射流雷诺数 *Re*，化学总当量比 Φ，几何涡旋数 S' 和喷嘴到目标距离 H/d。本课题用实验方法研究旋转 IDF 的火焰特征，包括火焰外观、火焰结构、火焰温度、火焰内气态物质、火焰总体污染物排放和火焰冲击传热。

旋转 IDF 外形像桃子。火焰结构分析将其分为三个特征区域。区域 1 是中心位置的回流区（IRZ），它靠近燃烧器出口。IRZ 回流区是由旋转射流中轴向逆压梯度引致，其体积占到火焰总体积主体。区域 2 是靠近下部的火焰边界区域，该区域一方面与外侧环境空气接触，另一方面在内侧与区域 1 接触。3 区则是靠近上部的火焰边界区域。IRZ 的 1 区通过具有逆向回流而与其他两个区域不同。在涡流旋转作用下，从燃烧器出来的流体颗粒依次通过区域 2 和区域 3，然后区域 3 中大部分流体颗粒逆转流动方向发生回流，流向燃烧器轴线及出口，形成 IRZ 的 1 区。因此，区域 3 可作为 IRZ 回流流体颗粒来源区域。

温度和火焰内部 O_2/CO/CO_2/NOx 浓度测量表明，这些参数的分布与流场耦合。可以看出，区域 1 是大体积、高温的 IRZ。由燃烧器供应的空气 / 燃料、燃烧产物和夹带的环境空气之间发生强烈混合，从而导致温度和气态物质浓度分布均匀。区域 2 在温度和气态物质浓度方面均具有明显梯度，这是因为在这个小体积区域内流场参数发生迅速变化。区域 2 颜色维持在海军蓝色不变，这是因供应的空气 / 燃料在这里发生强烈混合并同时发生强烈燃烧，这也表明该区域是火焰前沿面。区域 3 是后燃烧区，内部发生着燃烧产物的氧化、积聚和稀释，并且经过前两个区域后未烧尽的燃料在这里扩散燃烧。从 *Re* 和 Φ 对火焰内部气态物质浓度的影响可以看出，浓度分布与燃烧条件紧密耦合。其中，CO_2 和 NO_x 浓度具有与火焰温度相似的变化趋势，所以因热生成 NO 的机制会主导 NO_x 形成。从尾气测量中获得的数据显示，在某些燃烧条件下，中等水平 NO_x 排放和超低 CO 排放水平可以共存。

当旋转 IDF 向上垂直冲击到平坦表面时，旋转效应以三种方式影响局部热通量：1. 停滞点处的热传递被严重抑制；2. 局部热通量峰值发生在远离停滞点一定径向距离处；3. 局部热通量峰值的径向位置随着 H（喷嘴 – 目标距离）增加而更加偏离停滞点。存在一个最佳 H 值使得目标表面的热传递达到最大值，并且该最佳 H 在雷诺数和涡旋数不变时随着 Φ 增加而增加。

对旋转 IDF 与相同工况下非旋转 IDF 比较发现，旋转 IDF 火焰长度更短、更稳定。还发现，旋转 IDF 中 IRZ 的存在对火焰稳定性提高和长度缩短起到非常重要作用。相比于非旋转 IDF，由于有更强烈混合和更充分燃烧，旋转 IDF 在本质上是部分预混合火焰，因此它更清洁，烟灰形成较少。此外，旋转 IDF 较非旋转 IDF 产生略高 NOx 排放指数（EINOx）和更低的 CO 排放指数（EICO）。冲击热传递的比较表明，在非旋转 IDF 中产生冷核的小 H 位置旋转 IDF 具有更充分的燃烧，从而具有更高的传热速率。然而，旋转 IDF 在较高 H 处具有较低的热传递速率，而非旋转 IDF 此时能获得较充分燃烧从而具有较高热传递速率。这是因为在高 H 时旋转 IDF 火焰已经被夹带的环境空气充分冷却了。通过在相同的 Re 和 Φ、以及它们各自最佳 H 时比较旋转和非旋转 IDF，结果发现涡旋流对火焰总体热传递有负面影响。

在相同供气和送风条件下，比较了两种不同的燃气 / 空气混合机制，即预混（PMF）火焰和扩散（IDF）火焰。结果表明，两种旋转火焰具有相似视觉特征，包括火焰形状，尺寸和结构。这是因为作为影响旋转射流空气动力特性的重要参数：雷诺数和涡旋强度在两种火焰中几乎相同。PMF 和 IDF 两种火焰均被 IRZ 提升了稳定性，同时 IDF 的稳定性高于 PMF。这证实了 IDF 是预混火焰和扩散火焰的组合，并可以拥有两种火焰的优点。两种火焰中由燃烧器供应的空气 / 燃料、燃烧产物和夹带的环境空气之间保持了良好的混合，使得 IRZ 内气态物质包括 O_2、CO、CO_2 和 NO_x 在内的气体成分和温度分布均匀。与 IDF 比较，PMF 可获得较低水平的 NOx 和 CO 排放。在 PMF 中，燃料和空气的预混合显著增强了未燃混合物各向均匀性，并降低了燃烧区内燃烧生成物的停滞时间，从而降低了 NO_x 形成。此外，由贫燃燃烧产生的低温也有助于减少通过 NO 热机制形成的 NO_x 数量。PMF 具有较低 IDF, EICO 是由于燃料和空气的预混合以及较高的 O_2 浓度，这两个因素都促使 CO 向 CO_2 的转化。

CONTENTS

CHAPTER 1 INTRODUCTION

1.1 Combustion and flame

Combustion is the oldest technology of mankind and has furnished man with a major source of energy for more than one million years and at present, about 90% of our energy support (e.g., in traffic, electrical power generation, heating) is provided by combustion. The advent of nuclear energy has provided a rival energy source to combustion. However, many years will elapse before combustion loses its predominance and for the foreseeable future, it will continue to have important applications both as an energy source and in various industrial processes. Therefore, combustion will continue to be studied by researchers and quantitative understanding of this phenomenon is a desirable practical goal as well as of intrinsic interest.

Webster's Dictionary provides a definition of combustion as "rapid oxidation generating heat, or both light and heat; also, slow oxidation accompanied by relatively little heat and no light." This definition emphasizes the intrinsic importance of chemical reactions to combustion. It also emphasizes why combustion is so useful: combustion transforms energy stored in chemical bonds to heat which can be utilized in a variety of ways.

The objective of combustion is to retrieve energy from the burning of fuels in the most efficient way possible. Combustion can occur in either a flame or non-flame mode. What is a flame and how does it differ from other reacting systems? A flame is defined as a combustion reaction that can propagate subsonically through space. It is usually accompanied by the emission of visible radiation, a feature which is not essential to the definition. However, the property of spatial propagation is the important one which distinguishes flames from other combustion reactions. The spatial propagation of flames is a result of strong coupling between chemical reaction, the transport processes of mass diffusion and heat conduction, and fluid flow. Heat, active species, and radiation can all accelerate chemical reaction. Qualitatively, this can be considered as a positive feedback system. If the feedback exceeds some critical factor, the system will be self-sustaining. The limiting factor is the convection velocity carrying fresh material to the reaction, since the feedback is inversely proportional to it. The existence of flame movement implies that the reaction is confined to a zone which is small. This reaction zone is called the flame front, combustion wave, or combustion zone. We shall use the first designation in this thesis.

Flames are commonly classified according to three broad means: dispersion, aerodynamic flow, and initial physical state. The first of these has to do with the state of mixing of the reactants, or whether the reactants are premixed before entering the reaction zone. This criterion categorizes flames as being either premixed flames or non-premixed (diffusion) flames. In a premixed flame, the fuel and the oxidizer are mixed at the molecular level prior to the occurrence of any significant chemical reaction. Contrarily, in a diffusion flame, the reactants are initially separated, and reaction occurs only at the interface between the fuel and oxidizer, where mixing and reaction both take place. The term "diffusion" applies strictly to the molecular diffusion of chemical species, i.e., fuel molecules diffuse toward the flame from one direction while oxidizer molecules diffuse toward the flame from the opposite direction. In practical devices, both types of flames can be present in various degrees.

The second broad means of classification concerns the nature of the gas flow through the reaction zone in the fluid dynamic sense, i.e., whether it is laminar or turbulent. Laminar or streamlined flow implies that all mixing and transport must be done by molecular processes, while in turbulent flow this is aided by a macroscopic eddying motion. In turbulent non-premixed flames, turbulent convection mixes the fuel and oxidizer together on a macroscopic basis. Molecular mixing at small scales, i.e., molecular diffusion, then completes the mixing process so that chemical reactions can take place.

The third method of classifying flames specifies the initial physical state of the reactants—solid, liquid, or gas. Solid particle flames are probably best typified by those of coal dust in air. Liquid droplet or spray combustion is widely known in common oil burners, jet engines, etc., while gas flames are still more common and are self-explanatory.

These classes of flames—premixed or diffusion; laminar or turbulent; and gaseous, droplet, or particle—suffice as major divisions of flames. Other properties besides these serve to differentiate flames too, and the classification of a flame according to one of these subdivisions has an important bearing on its structure in one way or another, as will now be brought out.

Stationary flames or non-stationary flames

Deflagration or detonation flames

Open or enclosed flames

Normal or inverse flames

Stationary flames are opposed to non-stationary or propagating flames in the sense that the flames are stationary with respect to a reference point. The flame burning velocity is balanced by the flow velocity in a stationary flame. Most industrial flames belong to this category. Flame propagating down a tube filled with combustible mixture is an example of non-stationary flames.

The classification of flames into deflagration or detonation flames refers to the nature of flame burning velocity or flame speed. A deflagration flame propagates at subsonic velocity through gradual heat and mass transfer between the burned and unburned gas. A detonation flame is sustained by a shock wave through the high temperature and pressure rise behind the shock front. The flame front in a detonation flame propagates at super-sonic velocity.

The distinction between open or enclosed flames comes from the combustion system boundaries. Industrial flames usually occur in a vessel or combustion chamber with controlled fuel and air supplies. Such flames are called enclosed flames in contrast to open flames which are formed between a fuel jet and unbounded surrounding atmosphere. Flow patterns differ between an open and an enclosed flame since a vessel represents an impermissible mass boundary which affects fluid flow.

The last classification criterion considers the relative position of the fuel and oxidizer streams. In a normal flame, the fuel jet is surrounded by the oxidant jet. An inverse or reversed, reciprocal flame is defined here as a flame with a central oxidant jet surrounded by a fuel jet. In a broad sense, neither the fuel nor the oxidant needs to be a pure stream in an inverse flame. This criterion will be discussed in the next section.

1.2 Inverse diffusion flame

Inverse diffusion flames can be formed by two concentric tubes, comprising of an inner air jet and an outer fuel jet, under either confined condition or unconfined condition. Confined condition refers to an enclosed flame while an unconfined flame is an open flame in atmosphere. Since fuel and air supplies are initially separated, inverse diffusion flames are non-premixed in nature and the combustion performance is highly dependent on the degree of mixing between the fuel and air. A thin bell-shaped blue reaction zone will be formed at the jet exit under the equilibrium of jet velocity and flame burning velocity. The blue reaction zone is the surface of contact between the fuel and air where stoichiometric combustion takes place. In confined condition, however, it is very difficult to obtain a balance between jet velocity and flame burning velocity. Therefore confined inverse diffusion flames are usually unstable. In unconfined or open condition, fuel and air are injected into atmosphere. An air-fuel-air arrangement is formed, in which the central air jet is surrounded by the annular fuel jet and in turn the fuel jet is surrounded by ambient atmospheric air. When the central air jet velocity is low, the fuel jet will be in contact with the air jet at the inner side forming a bell-shaped blue flame and with the atmospheric air at the outer side forming an annular diffusion flame. When the central air jet velocity is high enough, the fuel jet will be entrained towards the air jet at the inner side and the mixing between the entrained fuel and the central air will produce a partially premixed flame. Even when the fuel is entrained by the central air and burned in premixed mode, part of the fuel may still burn in non-premixed mode due to either poor mixing or an excessive amount of fuel. This can result in a flame with upstream fuel-lean or stoichiometric combustion and downstream fuel-rich or non-premixed combustion. Such a flame configuration is similar to the staged combustion technique adopted for reducing NO_x emission through the process of NO_x-reburning. In staged combustion, usually a secondary air or fuel supply is provided to assist secondary combustion in the flue gas zone. The NO_x formed in the primary combustion zone is converted to HCN and eventually "burned away" in the secondary combustion zone. Inverse diffusion flames at high equivalence ratios will establish exactly similar staged combustion arrangement and therefore have potential application in low NO_x burners.

In recent years, inverse diffusion flames have gained popularity and attracted the attention of many researchers because they exhibit the characteristics between those of premixed and non-premixed flames. Due to the non-premixed nature, diffusion flames have a wide range of flammability even in turbulent state, but with a high soot-loading characteristic which sets a limit to their domestic applications where clean combustion is required. Premixed flames are cleaner and burn more intensely, but with a narrow range of operation due to the occurrence of flash-back and lift-off. With regard to inverse diffusion flames, by separately adjusting the fuel and air supplies, we can control the flame configuration from a premixed flame to a diffusion flame or a flame with characteristics in between.

1.3 Scope and objective of study

As a combination of premixed and diffusion flames, inverse diffusion flames are capable of exploiting the advantages of both premixed flames and diffusion flames, in regards to operational safety, pollutant emission, and flame stability. Specifically, inverse diffusion flames have no flash-back, reduced soot formation, a wide range of operational conditions and flexibility in flame length adjustment, coupled with potential NO_x-reburning capability. These advantages of inverse diffusion flames and their potential industrial and domestic applications have motivated this research work.

Numerous articles have been published dealing with the two basic flame types of premixed flames and diffusion flames, however, in the history of flame research, only a few investigations on inverse diffusion flames have been carried out. Among them, the study of flame structure is the specialized topic of interest to combustion scientists and engineers. Flame structure is the term used in one sense to describe the process taking place within the flame itself and in another sense to describe the external appearance of the flame. Besides flame structure, former investigations are mainly focused on their low soot-loading characteristic. Other concerns include the study of pollutant formation mechanisms and the application of inverse diffusion flames in flame impingement heat transfer.

Energy conservation and environmental concerns emphasize the need for fundamental investigation of the mechanisms of pollutant formation. So both thermal and emission characteristics of swirling inverse diffusion flames are to be investigated in details. This study also aims to better understand the phenomena such as flame extinction, quenching and heat and mass transport properties.

This dissertation aims at investigating and comparing inverse diffusion flames with and without swirl. The introduction of a swirling motion to the inverse diffusion flame is proposed to reduce its flame length. The governing parameters of the inverse diffusion flame with swirl will be fully identified and their effects on flame length, flame stability and preferential separation distance for heat transfer etc. will be investigated. It is the objective of this study to exploit the feasibility of utilizing swirling inverse diffusion flames in flame impingement heat transfer.

We begin our study of swirling inverse diffusion flames by developing a swirl burner and testing its performance. The burner aims to generate flames operating in high swirl mode. Then, detailed investigations go to the followings:

1. Observe the flame appearance in the full range of experimental conditions and identify the flame structure of the inverse diffusion flame. Observe the change in both flame appearance and structure; namely, investigate the effect of air jet Reynolds number and overall equivalence ratio on the flame appearance and structure.

2. Measure the temperature field of the inverse diffusion flame, and find out the influence of air jet Reynolds number and overall equivalence ratio on the flame temperature. Based on the data of flame temperature, study the combustion condition in the flame and study the relationship between the flame structure and the thermal characteristics of the inverse diffusion flame.

3. Probe the in-flame stable gaseous species of the inverse diffusion flame, since their concentrations help understand the combustion condition in the flame. Investigations are to be based on the effects of air jet Reynolds number and overall equivalence ratio on the species concentration distribution. The overall pollutant emissions characteristics of the flame will also be studied, with the focus on CO and NO_x emissions, because for LPG combustion, CO and NO_x are the major pollutants emitted. On the other hand, either CO or HC is a good indicator of incomplete combustion, but more attention should be paid on CO as slightly excessive CO emission can be fatal in domestic applications of the IDF.

4. Survey the heat transfer characteristics of the inverse diffusion flame upon impinging vertically onto a flat plate under different combination of the air jet Reynolds number, overall equivalence ratio and the nozzle-to-plate distance. Detailed studies will go to the effects of each of these parameters including air jet Reynolds number, overall equivalence ratio and the nozzle-to-plate distance, on the heat flux distribution along the target plate.

5. Under the same or similar experimental conditions, compare the thermal and emission characteristics of the inverse diffusion flame operating in high swirl mode and another inverse diffusion flame without induced swirl. Identify the effect of swirl on the flame appearance, flame structure, flame temperature and heat transfer characteristics. Comparison further goes to the swirling inverse diffusion flame and a swirling premixed flame. Both the similarity and distinction in thermal and emission characteristics of the swirling inverse diffusion flame and the swirling premixed flame will be found out, which will help us understand the features of these two different flame types.

CHAPTER 2 LITERATURE REVIEW

2.1 Review of normal diffusion flame

As stated in Chapter 1, diffusion flames have received far less attention than premixed flames. The lack of measurable physical parameters is one of the major reasons. The term "diffusion flame", in fact, was first applied by Burke and Schumann (1928) in their pioneering investigation into the structure of enclosed diffusion flames. Since refinements to the Burke and Schumann theory have been made by a number of investigators, the classical Burke and Schumann theory is reviewed in detail here.

In their study, a jet of gaseous fuel from a cylindrical tube was discharged into a cylindrical concentric tube where the oxidizer was flowing in the annulus. Two types of flames were observed: the one with the conical flame surface converging towards the tube axis was called an "over-ventilated" flame; another one with the shape of an upside down cone was called an "under-ventilated" flame. The term "diffusion" was used to describe the process that the oxidizer diffuses into the fuel and thus causing simultaneous mixing and combustion. The rate of combustion depends on the rate of mixing between the oxidizer and the fuel rather than the much faster chemical reaction rate. Due to insufficient supply of oxidizer in the core of the fuel stream, the combustion thereunto generates high concentration of unburned hydrogen, carbon monoxide and extremely high soot-loading products inside the flame. They predicted flame shape and length by solving a diffusion equation:

$$\frac{\partial C}{\partial Z}=\frac{D}{U}\left(\frac{\partial^2 C}{\partial r^2}+\frac{1}{r}\frac{\partial C}{\partial r}\right) \tag{2.1}$$

Where U is the axial gas velocity and D is the diffusion coefficient. Several assumptions have to be made to yield Equation 2.1 as a governing equation: (a) the fuel and air velocities are equal; (b) axial diffusion is negligible; (c) density and diffusity are constant throughout the flame; (d) flame structure is located at the stochiometric mixture surface; (e) infinite chemical reaction rate exists such that the flame front is a geometric surface; (f) no thermal or pressure mass diffusion exists.

Barr (1953) investigated the appearance of enclosed, cylindrical butane flames. He characterized ten different types of normal diffusion flames phenomenologically from his observation. It was found that over-ventilated and under-ventilated flames could only be obtained under proper fuel and air flow rates.

The height of diffusion flames has been the concern of many researchers since it is the only easily measurable parameter. Barr (1954) studied the length of enclosed laminar butane-air flames experimentally. The influences of burner tube diameter and tube thickness, as well as fuel and air flow rates were discussed. Savage (1962) discussed the Burke and Schumann model from a fluid-dynamic point of view. He presented a modified equation for enclosed laminar diffusion flame heights. Ropper (1977) modified the Burke and Schumann model so as to satisfy the discontinuity condition between fuel and air velocities. His theory was tested both in a circular port burner and a slot burner. He concluded that for non-axisymmetric burner, diffusion flame height might be controlled either by the momentum of the fuel jet or by buoyancy effects.

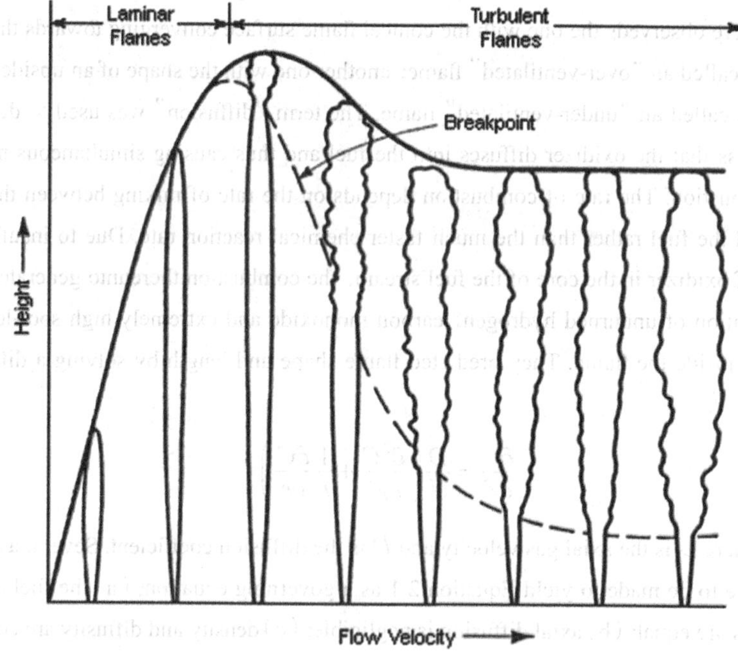

Figure 2.1 Change in flame height with increase in How velocity (Hottel and Hawthorne 1949).

Hottel and Hawthorne (1949) observed changes in flame shape by progressively increasing the flow rate of fuel gas issuing from a round nozzle. They found that the length of a diffusion flame is dependent on the fuel jet velocity and the associated turbulence. When fuel gas was discharged at velocities below a critical value from the nozzle into stagnant air surroundings, the flow of gas is laminar and the mixing of gas and air occurs by molecular diffusion in a thin flame surface which is fixed in space. As the fuel jet velocity was increased, the diffusion flames increase in length until a critical velocity is reached and the tip of the flame becomes unsteady and begins to flutter. With further increase in velocity, this unsteadiness develops into a noisy turbulent brush of flame starting at a definite point along the flame where the laminar flow breaks down and a turbulent jet flow develops. The distance from the nozzle to the point where the turbulent brush begins is termed the breakpoint length. Characteristic flame length and breakpoint length curves as initially determined by Hottel and Hawthorne are shown in Figure.2.1. As the fuel jet velocity increases from zero, initially, there is an almost proportional increase in flame length and, at any velocity in this region, the flame is sharp edged and constant in shape. When a critical Reynolds number of 2000 is exceeded, the tip of the flame changes in character and a slight brush forms. With further increase in velocity, the breakpoint moves towards the nozzle and, with some fuels, the flame length is slightly reduced. When the breakpoint has approached quite close to the nozzle, the fully developed turbulent flame condition is formed. Further increase in velocity has practically no effect on flame length, but flame noise continues to increase and flame luminosity continues to decrease. Finally, the flame blows off the nozzle. Feikema *et al.* (1991) reported that a co-flowing stream of oxidizer also leaded to the same flame shortening effect due to enhanced turbulent mixing between the co-axial streams of fuel and air.

Wolfhard and Parker (1949) studied the reaction zone structure of H_2-O_2, NH_3-O_2, CH_4-O_2 flames in a flat flame burner spectroscopically. For a CH_4-O_2 flame, the results showed that oxygen does not reach to the luminous zone but that a strong OH band occurs on the oxygen side. Their efforts advanced the knowledge of the structure of the very thin diffusion flame zones. Smith and Gordan (1956) measured stable gaseous product concentrations from various regions of a methane diffusion flame with a mass spectrometer. They found that methane is first pyrolyzed in the inner mantle of the flame and the pyrolysis products are then oxidized in the high temperature reaction zone. In the investigation of flame temperature of diffusion flames, Becker and Yamazaki (1978) measured the temperature distributions of turbulent diffusion flame jets without co-flowing air. They found that due to the mechanism of diffusion, the peak temperature or the primary reaction zone of the flame usually occurs at the air/fuel interface. Similar results were published by Lockwood and Moneib (1982) . Mitchell *et al.* (1980) measured temperature, velocity and stable gaseous product profiles from an enclosed normal diffusion flame. They reported that the primary reaction zone of a diffusion flame is nominally 2 mm thick and the luminous flame surface marks the fuel-rich side of the reaction zone and the fuel-lean side is marked by the burnout of CO species. Their results also showed that the concentration and temperature profiles could be generalized by plotting them as a function of local equivalence ratio.

Soot is one of the major pollutants formed in diffusion flames and its emission from the flame depends on the competition between the soot formation and soot oxidization processes. Wolfhard and Parker (1949) suggested that soot particles could not be burned in the oxygen but probably react with H_2O, or OH. Yagi and Iino (1962) investigated the soot formation characteristics of diffusion flames and pointed out that soot particles formed in the flame assist radiation heat transfer. Echigo *et al.* (1967) reported similar results. Sidebotham and Glassman (1992) revealed that in a specified experimental system, a diffusion flame tends to start producing soot at different heights or different fuel flow rates.

Numerical studies have also been carried out in the analysis of diffusion flames. Mitchell (1980), based on the Burke and Schumann model, developed a modified numerical model on diffusion flames which is capable of predicting the effects of flow rate, equivalence ratio, preheat, and nozzle size on the thermal and aerodynamic fields established in confined, laminar diffusion flames. Peters (1986) and Chou *et al.* (1998) used the flamelet approach to develop a numerical model for NO formation in laminar Bunsen flames. The flamelet approach can be considered as the extension of the Burke and Schumann model and it assumes infinitely fast chemical reaction such that the reaction zone is an infinitely thin interface. Under constant pressure combustion without heat loss and equal diffusivity assumptions, the thermo-chemical properties are determined completely by the local mixing state known as the mixture fraction. By introducing the scalar dissipation rate as a parameter to describe the degree of departure from the equilibrium state, Peters successfully modeled the formation of thermal NO in his study and suggested that NO is generated near the reaction zone.

2.2 Review of inverse diffusion flame

Normal diffusion flames are a type of diffusion flame with an inner fuel jet surrounded by an outer air jet. Inverse diffusion flames are similar to normal diffusion flames, except that the relative positions of fuel and air are reversed. While there is fair amount of information on normal diffusion flames, reported work on inverse diffusion flames is rare.

Friend (1922) appeared to be the first person to show the concept of inverse diffusion flame. In his monograph, he described "inverse diffusion flame" as "reciprocal flame" . Burke and Schuman (1928) measured the height of inverse carbon monoxide-air flames and claimed that the agreement between their experimental and predicted flame heights was good. The first comprehensive study on inverse diffusion flames was performed by Wu (1984) under confined condition. The inverse diffusion flame investigated by Wu is a blue bell-shaped laminar flame attached to the air jet exit. At high overall equivalence ratio, an orange-yellow "cap" is developed upon the blue reaction zone. In between the orange-yellow "cap" is a dark zone. Wu and Essenhigh (1984) mapped out six different types of inverse diffusion flames according to their different flame appearance and stability. They compared, for laminar normal diffusion flames and inverse diffusion flames, the axial and radial species concentrations and the temperature profiles. Results indicated that the peak temperature of the inverse diffusion flame is reasonably coincident with the blue reaction zone and symmetrical along the center axis. However the inverse diffusion flame reported is a laminar flame, and high air jet velocity leads to flame blow-off on the burner.

Takagi *et al.* (1996) studied the difference between normal and inverse diffusion flames, using nitrogen diluted hydrogen as the fuel. The species concentrations were considered in both of the flames to predict the temperature. They found that the flame temperature of the inverse diffusion flame is higher than that of the corresponding normal diffusion flame at the same flame height.

Huang *et al.* (1997) carried out similar research on double concentric jets consisting of a central air jet and an annular fuel jet. The flame was subdivided into four types at low air jet velocities and five types at high air jet velocities. In the low air jet velocity range, a central blue flame always exists and is stabilized on the central air jet exit. The central flame may eventually lift off at higher fuel jet velocity. Due to the lengthened mixing distance between the fuel and air the lifted central flame may have characteristics of a partially premixed flame.

Wentzell (1998) investigated the inverse diffusion flame in unconfined condition. He measured the flame temperature and flame length at different fuel and air jet diameters and at different fuel and air Reynolds numbers. The inverse diffusion flame studied shows a dual flame structure composed of a premixed region located in the upstream portion of the flame and a diffusion flame downstream. He discovered that for this kind of flame, diffusion is not the main transport mechanism and thus suggested that the flame could be classified as a partially premixed flame, and the degree of partial premixing depends on the nozzle geometry and the flow conditions.

Mikofski *et al.* (2006) measured the flame length of the inverse diffusion flame by examining the position of planar laser-induced fluorescence of hydroxyl radicals, and it was found that inverse diffusion flames are similar in structure to normal diffusion flames. Bindar and Irawan (2002) examined the shape of the inverse diffusion flame at high level of fuel excess and found that the flame shape is different from those of Wu and his co-worker (1984). They proposed that the shape of the inverse diffusion flame is affected by both the inlet air momentum and the inlet fuel momentum. At low inlet air velocity, the flame shape is similar to that of a normal diffusion flame. At higher air velocity, usually 5–16 times higher than the fuel velocity, the flame shape is composed of two regions: a base and a tower on top of the base. Sobiesiak and Wentzell (2005) examined the flame centerline mean temperature, temperature fluctuation, and the Schlieren image, using jets of different diameters. The temperature profiles together with the Schlieren images showed the presence of partial premixing zone in the inverse diffusion flame. The well-mixed reaction zone usually appears in the flame center, enveloped by an outer diffusion flame.

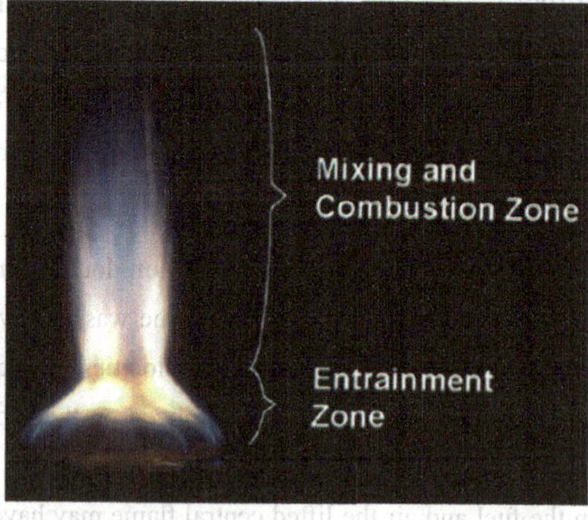

Figure 2.2 CAP flame at *Re*=2500 and *Φ*=1.0 (Sze *et al.*2004, 2006).

Sze *et al.* (2004, 2006) developed a burner with twelve circumferentially arranged fuel ports surrounding a central air port (CAP), different from the arrangement with one annular co-axial fuel port surrounding a central air port (CoA). They reported that the CAP flame can be divided into two zones separated by a neck: a lower entrainment zone, and an upper mixing and combustion zone, as shown in Figure.2.2. A bluish fluctuating reaction zone is formed in the mixing and combustion zone. In the CoA flame, fuel entrainment is obvious only at high air jet Reynolds numbers and normally the air jet co-flows with the fuel jet. The flame temperature measurements revealed that at higher overall equivalence ratio, the maximum temperature in the CAP flame is higher than that in the CoA flame. Moreover, in both CAP and CoA flames, a cool core exists at low flame heights and disappears at high flame heights where the maximum temperature zone occurs in the flame center and drops steadily away from the centerline.

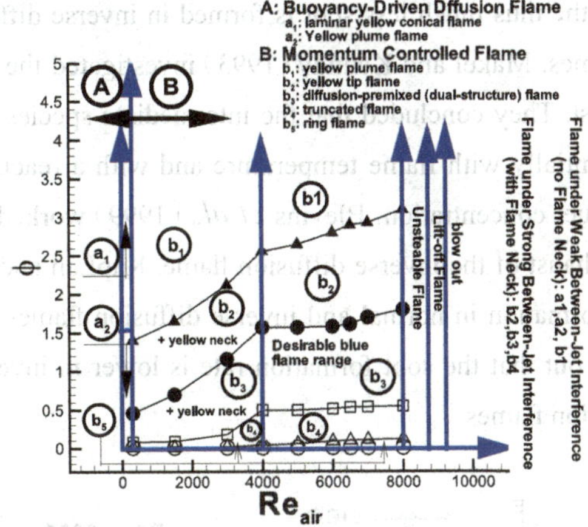

Figure 2.3 Mapping of flame structure based on *Re* and *Φ* (Dong *et al.*2007).

Similar investigations were made by Dong *et al.* (2007) on the flow pattern, flame structure and pollutants emission characteristics of a CAP inverse diffusion flame. They concluded that the flame shape and structure are mainly dependent on the air jet velocity and the ratio of fuel and air velocities, and that the air jet Reynolds number and the ratio of the velocities of the air to fuel jet are two key factors influencing the level of the between-jet interference. Based on the critical air jet Reynolds number and critical overall equivalence ratio, a total of seven different flame structures were identified: buoyancy-driven yellow tip flame, buoyancy-driven yellow plume flame, ring flame, open flame, dual-structure triple-layer flame, momentum-controlled yellow tip flame and momentum-controlled yellow plume flame. The dual-structure triple-layer flame is in the color of blue and consists of a base diffusion flame and a flame torch separated by a flame neck. The flame torch resembles a premixed flame with an inner reaction zone and an outer flame layer. This type of inverse diffusion flame has the advantage of no flashback, high flame temperature and acceptable pollutants emission level.

Many studies have been carried out on the low soot loading in inverse diffusion flames. As the inverse diffusion flame is non-luminous and the normal diffusion flame is luminous, the radiant emissions are significantly different between these two flame types. Soot particles do not experience the oxidation process in inverse diffusion flames and thus they are generally considered ideal for studying incipient soot formation. Sidebotham and Glassman (1992), studied the soot formation characteristic of inverse diffusion flames and normal diffusion flames under the effect of temperature, fuel structure and fuel concentration and discovered that the surface growth rate of inverse diffusion flames is very small along the particle path, thus much less soot is formed in inverse diffusion flames than in normal diffusion flames. Makel and Kenndy (1993) investigated the carbon atoms in the flame and the exhaust. They concluded that the intermediate species concentration in the flame scales exponentially with flame temperature and with a reaction order near unity with respect to the fuel concentration. Blevins *et al.* (1999) worked on the existence of young soot in the exhaust of the inverse diffusion flame. KapLan and Dailasanath (2001) compared the soot formation in normal and inverse diffusion flames under the flow-field effect. They pointed out that the soot formation rate is lower in inverse diffusion flames than in normal diffusion flames.

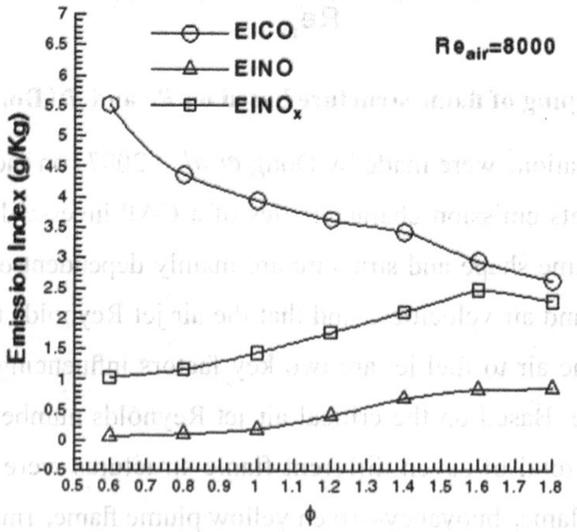

Figure 2.4 Emission indices of pollutant emissions (Dong *et al.*2007) .

Another critical pollutant of concern is nitrogen oxides emitted during the combustion of inverse diffusion flames because of their roles in the formation of both photochemical smog and acid rain. William *et al.* (1995) reported the $EINO_x$ of a staged-air burner which involves inverse diffusion combustion. The purpose of applying inverse diffusion flames in the staged burner is for low NO_x emission, in which air is injected into the fuel-rich combustion products. They concluded that the residence time is a major factor influencing NO formation in inverse diffusion flames and that the mixing between the primary products and secondary air is a key parameter in NO_x reduction and the NO_x emission levels off once turbulent mixing is established. Fleck (1998) also investigated the inverse diffusion flame formed in a staged-air burner. The high velocity air jet entrains the fuel from the surrounding fuel jets, which are diluted by the surrounding flue gas. Because of the dilution effect, the burner generates a large combustion zone of relatively low temperature that is unfavorable to the formation of NO_x. William *et al.* (1999) reported both the flame temperature and the in-flame NO concentration. It was shown that most of the inverse diffusion flame attributable NO is generated at the tip of the flame. In the work of Sze *et al.* (2006), it was shown that at fixed air jet Reynolds number of 2500 and for the overall equivalence ratio of 1.0 to 6.0, the $EINO_x$ curve for both CAP and CoA flames is bell-shaped, with a maximum value of 3.2 g/kg at the overall equivalence ratio of 1.2 for the CAP flame and 3.0 g/kg at the overall equivalence ratio of 2.2 for the CoA flame. Dong *et al.* (2007) conducted emission measurements on dilution air-free basis. They found that at fixed air jet Reynolds number of 8000, the EICO decreases with the increase in overall equivalence ratio, while the EINO and $EINO_x$ follow the inverse trend. The inverse diffusion flame in their study emits a comparable level of NO and NO_x to those from partially premixed flame and staged-air burner.

The feasibility of inverse diffusion flames in the application of flame impingement heat transfer has been investigated by several researchers. Sze *et al.* (2006) studied the heat transfer characteristic of the CAP inverse diffusion flame upon impinging on a flat plate under different combinations of the air flow rate, the fuel flow rate and the nozzle-to-plate distance. They revealed that the heat flux is greatly affected by three factors: air jet Reynolds number, overall equivalence ratio and nozzle-to-plate distance. At fixed nozzle-to-plate distance, both Reynolds number and overall equivalence ratio should be adjusted for the highest stagnation point heat flux to occur. With fixed Reynolds number, there is an optimal nozzle-to-plate distance and an optimal overall equivalence ratio for improving radial heat flux distribution. In a comparison with the premixed flame, they showed that the inverse diffusion flame has lower peak heat flux but can provide more uniform heating around the stagnation point. Ng *et al.* (2007) performed experimental studies on the heat transfer characteristics of both an inverse diffusion flame and a premixed flame. The effects of Reynolds number, overall equivalence ratio and nozzle-to-plate distance on the heat flux in the two flames were investigated and compared. The area-averaged heat flux and the heat transfer efficiency were calculated from the radial heat flux distributions. For the inverse diffusion flame, the area-averaged heat flux increases with either Reynolds number or overall equivalence ratio, because both imply an increase in the fueling rate. The comparison with the premixed flame showed that both the stagnation point heat flux and the radial heat flux distribution are quite different. The premixed flame has intense combustion in a shorter flame length range while the inverse diffusion flame has intense combustion over longer flame length. Furthermore, the inverse diffusion flame has comparable or even higher heat transfer efficiency than the premixed flame over a large range of nozzle-to-plate distance. In the work of Dong *et al.* (2007), they pointed out that among the seven flame types identified in unconfined condition, the dual-structure triple-layer flame combines the advantages of both diffusion and premixed combustion and has the feasibility for domestic heating applications. Therefore, the flame structure, static wall pressure characteristics, and local and averaged heat transfer characteristics of this type of inverse diffusion flame were examined. By increasing the nozzle-to-plate distance gradually, the flame structure varies from the impingement of the cold gas inside the inner reaction zone, to the impingement of the tip of the reaction zone, and finally to the impingement of the post flame zone. The highest heat transfer occurs when the tip of the inner reaction zone impinges on the plate. The heat transfer rate from the impinging inverse diffusion flame is higher than that from the impinging premixed flame due to the augmented turbulence level originated from the flame neck.

There are also a small number of other special studies on inverse diffusion flames. Hwang and Gore (2002) found that the inverse methane/oxygen flame is more effective in generating a strong radiation heat flux compared to the normal methane/oxygen flame, because the former produces more soot. Sunderland *et al.* (2004) studied the effects of oxygen enhancement and gravity on normal and inverse diffusion flames. Lee *et al.* (2004) applied an ethylene inverse diffusion flame for the synthesis of carbon nano-tubes on a catalytic metal substrate.

2.3 Review of swirl combustion

The dramatic effects of swirl in inert and reacting flow systems have been known and appreciated for many years. Some effects are favorable and designer strives to generate the required amount of swirl for this particular purpose; other effects are undesirable and the designer is then at pains to control and curtail their occurrence. In combustion systems, the strong favorable effects of applying swirl to injected fuel and air are extensively used as an aid to stabilize the high intensity combustion process and efficient clean combustion in a variety of practical situations, such as swirl combustors, cyclone combustors, and swirl burners. Swirl provides an effective aerodynamic means to stabilize the flame, control the combustion process, enhance the heat and mass transfer, and increase the combustion efficiency.

Swirling flows result from the application of a spiraling motion, a swirl velocity component being imparted to the flow by the use of swirl vanes, by the use of axial-plus-tangential entry swirl generators or by direct tangential entry into the chamber. The degree of swirl is usually characterized by the swirl number *S*, which is a non-dimensional number representing axial flux of swirl momentum divided by axial flux of axial momentum, times the equivalent nozzle radius:

$$S = \frac{G_\varphi}{G_x R} \tag{2.2}$$

$$G_\varphi = \int_0^R W \rho U 2\pi r^2 dr \tag{2.3}$$

$$G_x = \int_0^R \rho U^2 2\pi r dr + \int_0^R p 2\pi r dr \tag{2.4}$$

where U, W and p are the axial and tangential components of velocity and static pressure, respectively. In a free jet in stagnant surroundings, both G_x and G_θ are constants.

The effects of inlet flow swirl on the subsequent flow field produced are increasingly dramatic as the degree of swirl increases. The effect of a low degree of swirl (weak swirl, $S<0.4$) is to increase the width of a free or confined jet flow: jet growth, entrainment and decay are enhanced progressively as the degree of swirl is increased. Although there may be significant lateral or radial pressure gradients, they do not give rise to more than a slight longitudinal or axial pressure gradient. At higher degrees of swirl (strong swirl, $S>0.6$), strong radial and axial pressure gradients are set up near the nozzle exit, resulting> in axial recirculation in the form of a central toroidal recirculation zone, which is not observed at weaker degrees of swirl. Of course, the precise effect is found to depend on many factors as well as the swirl number; for example, nozzle geometry, size of enclosure, if any, and the particular exit velocity profiles.

An extensive review on combustion in swirling flows was conducted by Syred and Beer (1974). They pointed out that the most significant effects of swirling flow are produced by recirculation and thus their paper was mainly concerned with swirling flows associated with recirculation. Initially, the isothermal performance of swirl combustors was considered, and it was demonstrated that the flow is often not axisymmetric but three-dimensional time-dependent. Over the many different types of swirl generators, they found that despite the difference in configuration, there are many similarities in flow patterns: a large toroidal recirculation zone is formed in the burner exit; very high levels of kinetic energy of turbulence are formed just by the exit lip, with a rapid decay occurring after on exit diameter. It was shown that these high levels of turbulence are caused, in fact, by a three-dimensional time-dependent instability of swirling flows, named precessing vortex core (PVC). Under most non-premixed combustion conditions, the swirling flow returns to axisymmetry, because combustion has a decisive effect on the PVC and thus on the aerodynamics of the flow produced by swirl burners. The general temperature characteristics of a swirling flame are: maximum temperatures are located near the burner exit just outside the boundary of the reverse flow zone; temperature levels throughout the reverse flow zone are nearly uniform; sharp temperature profiles and gradients are found in the vicinity of the reaction zone within the flame.

Tangirala *et al.* (1987) quantified the effects of heat release and swirl on the internal recirculation zone, mixing, velocity probability density functions and flame blow-off limits. It was found that increasing the heat release by changing the overall stoichiometry results in benefits such as an increase in the recirculation. The effect of increasing swirl, stated by Feikema *et al.* (1990), is improve the mixing and flame stability for swirl numbers up to one. If sufficient swirl is imparted ($S>0.6$) to create a recirculation zone, up to a fivefold improvement in flame stability occurs. They also stated that further increase of swirl actually reduces the turbulence level and flame stability. In addition, excessive swirl has the disadvantage of forcing the flame to move upstream into the quarl due to increased adverse pressure gradients. Cha *et al.* (1999) studied the effect of swirl on lifted flames in a non-premixed jet by rotating a nozzle to generate a continuous swirl. Results indicated that the lift off height decreases linearly with nozzle rotational speed in turbulent jets.

Terasaki (1996) achieved low NO_x emission by a non-premixed, direct fuel injection burner equipped with a double swirler. By means of strong swirling flows, the mixing of fuel and air in the combustion region is enhanced to increase mixture homogeneity and shorten the characteristic time for NO_x emission. In an investigation of flame structure and NO_x emission of swirling methane jet flames, Cheng *et al.* (1998) modified the initial fuel-air mixing ahead of the flame by means of the swirl number, the fuel-air momentum flux ratio and the fuel injection location. Laser Doppler velocimetry (LDV) measurements showed that the shear layer between the edge of the internal recirculation zone and the external flow is a highly turbulent and rapid mixing region. The maximum mean flame temperature is located in this region, indicating violent combustion and strong mixing of fuel, air and hot combustion products. Strong and rapid mixing increases mixture homogeneity and shortens the characteristic time for NO_x formation. Schmittel *et al.* (2000) carried out an experimental investigation of the flow field and NO_x formation in turbulent swirling flames. They concluded that swirl-stabilized non-premixed flames offer the possibility to control the flow and mixing pattern and thus improve ignition stability and minimize NO_x emission by adapting the aerodynamic inlet conditions.

Olivani *et al.* (2007) obtained a comprehensive picture of the near field of a swirl-stabilized non-premixed flame under both isothermal and reactive conditions. The experimental results indicated that the method of fuel injection plays an important role on mixture formation and flame stabilization.

Compared with investigations on high swirl combustion, there are only a few papers on low swirl combustion. Chan *et al.* (1992) studied a novel weak swirl burner for premixed flames under a broad range of mixture and turbulence conditions. The form of the low swirl burner (LSB) may show similarity to that of a conventional high swirl burner, but the LSB functions quite differently. The LSB exploits the propagating nature of a premixed turbulent flame and does not rely on flow recirculation to anchor the flame. The stabilization mechanism is similar to that of stagnation flow stabilized flames. Weak swirl generates a divergent flow above the burner where the mean axial velocity decreases almost linearly. This provides a very stable configuration for the flame to maintain itself at the position where the local mass flux equals the burning rate. Even under very intense turbulence, it stabilizes lean premixed flames very close to the flammability limit. Cheng *et al.* (2000) scaled the original 5.28 cm inner diameter LSB to 10.26 cm inner diameter and investigated up to a firing rate of 586 KW, with the purpose to demonstrate its viability for commercial and industrial furnaces and boilers. Zhao *et al.* (2004) conducted a numerical simulation of an open swirl-stabilized turbulent premixed flame. One of the flames, SWF4, as reported by Chan *et al.* (1992) was modeled. They revealed that flame stabilization relies on flow divergence instead of recirculation although a recirculation zone is located downstream of the flame.

There have been some published papers on the heat transfer characteristics of single and multiple impinging swirling isothermal jets. Ward and Mahmood (1982) used the naphthalene sublimation technique to investigate the effect of heat and mass transfer on a flat surface from a swirling impinging jet. They found that the measured radial distribution of local Nusselt number is slightly more uniform than for a conventional impinging jet, but its values are significantly lower, particularly in the vicinity of the stagnation point. They concluded that the swirl has an unfavorable effect on heat transfer in terms of both local and average value of the parameters tested. The same authors (1993) examined the effect of swirl in a rectangular nozzle array. In this case, interaction between neighboring co-rotating jets leads to an overall heat transfer enhancement. Such an enhancement was observed only at swirl numbers of 0.12 to 0.24 when the nozzle array is relatively close to the impingement surface. Huang and El-Genk (1997) used a swirl generator made of a cylindrical plug with four narrow channels to provide swirl to single and multiple impinging air jets. Their results demonstrated improvement in the radial uniformity and large increases in the local and average Nusselt number. Owsenek *et al.* (1997) carried out numerical investigations of impinging jet with superimposed swirls. However, their study was limited to the laminar flow regime. Lee *et al.* (2002) tested on a vane-type swirl generator and measured the local Nusselt number for a swirling round turbulent jet impinging on a flat plate. They found that the effect of swirling jet flow is mainly significant near the stagnation region and for small nozzle-to-plate distances. The average Nusselt number is larger than that with non-swirling flow. But for large nozzle-to-plate distances, the effect of swirling jet flow is rarely seen and the average Nusselt number can be lower than that with non-swirling flow. The experimental results from Yuan *et al.* (2006) revealed that the swirling jet causes change of the heat transfer distribution significantly. The heat transfer in the stagnation region deteriorates to some extent and rises in the wall jet region, thus the radial uniformity is improved.

Only a few studies of the heat transfer characteristics of single and multiple impinging swirling flame jets have been made. Huang *et al.* (2006) showed that a circular laminar premixed flame jet with induced swirl has more uniform heat flux distribution and higher temperature in comparison with the flame jet without induced swirl. An array of three identical laminar premixed flame jets with induced swirl was developed by Zhao *et al.* (2008), who suggested that under low pressure and low Reynolds number condition, a swirling flame jet array can achieve more complete combustion at shorter nozzle-to-plate distance and provides more uniform heat transfer on the target surface. Furthermore, the heating efficiency is enhanced when induced-swirl is applied. The author could not find any earlier studies on the heat transfer characteristics of impinging swirling flames operating in high swirl mode which involves flow recirculation, even though there have been some studies on the heat transfer characteristics of impinging flames with low swirl.

2.4 Summary

The inverse diffusion flame has distinctly different properties from the normal diffusion flame. The inverse diffusion flame is capable of obtaining the advantages of both premixed flame and diffusion flame. As far as pollutant emissions are concerned, the inverse diffusion flame has the potential of reduced soot formation and low NO_x emission. The controllable flame length and flame structure characteristics together with no danger of flash back broaden the potential application of inverse diffusion flames.

All previous studies on inverse diffusion flames have enriched our knowledge of their characteristics. However, most researchers performed their work on inverse diffusion flames without swirl, and there has been no investigation on swirling inverse diffusion flames yet. Inverse diffusion flames without swirl can be very long when overall equivalence ratio is high. For certain applications especially domestic ones, it is necessary to shorten the flame and simultaneously broaden its effective contact area with the heating target for better heat transfer.

The benefits of swirl combustion depend in most instances on the formation of an internal recirculation zone, which allows flame stabilization to occur in regions of relative low velocity where the flow velocity and the flame speed can be matched, aided by the recirculation of heat and active chemical species. Swirl can improve flame stability as a result of the formation of toroidal recirculation zones and to reduce flame lengths by producing high rates of entrainment of the ambient fluid and fast mixing.

Most previous studies on impingement heat transfer were carried out on laminar jets or premixed flame jets without swirl or with low-swirl which does not involve flow recirculation. The feasibility of using a swirling inverse diffusion flame with induced high swirl has not been touched on up to date.

As a result of the literature review, the present research aims to develop a burner to operate an inverse diffusion flame in high swirl mode. The flame length of such an inverse diffusion flame is expected to be reduced by the induced high swirl. The flame stability is expected to be enhanced by the internal recirculation zone. The soot formation and NO_x emission are expected to be lowered by fast mixing of fuel and air occurring in the internal recirculation zone. To better understand the effect of swirl on the thermal and emission characteristics of flames, comparison of different inverse diffusion flames operating under identical or similar experimental conditions will be conducted. Investigation of the impingement heat transfer characteristics of the swirling inverse diffusion flame associated with flow recirculation is also a main objective of this study. The influences of governing operational parameters including Reynolds number, overall equivalence ratio and nozzle-to-plate distance on the heat transfer will be studied in detail.

CHAPTER 3 EXPERMENTAL APPA-RATUS

3.1 Introduction

This chapter presents the experimental rigs, apparatus and measuring techniques used in the investigation of inverse diffusion flames (IDFs) with induced swirl. For the purpose of clarity, the whole system of experiments is divided into two groups. The first group is related to the investigation of the characteristics of the open flame, including the flame appearance, flame length, flame structure, flame temperature, in-flame stable gaseous species and overall pollutant emissions. The second group is concerned with the investigation of the characteristics of the impinging flame, including the flame temperature, flame visualization, wall static pressure and heat transfer characteristics. The experimental set-up for each group of experiments is shown in Figures 3.1 and 3.2, respectively.

All the experiments were conducted at the Heat Transfer & Combustion Lab which is located on the ground floor. The room is well ventilated and the temperature is kept at 21℃ by air-conditioners. The flames studied operate in atmosphere without confinement.

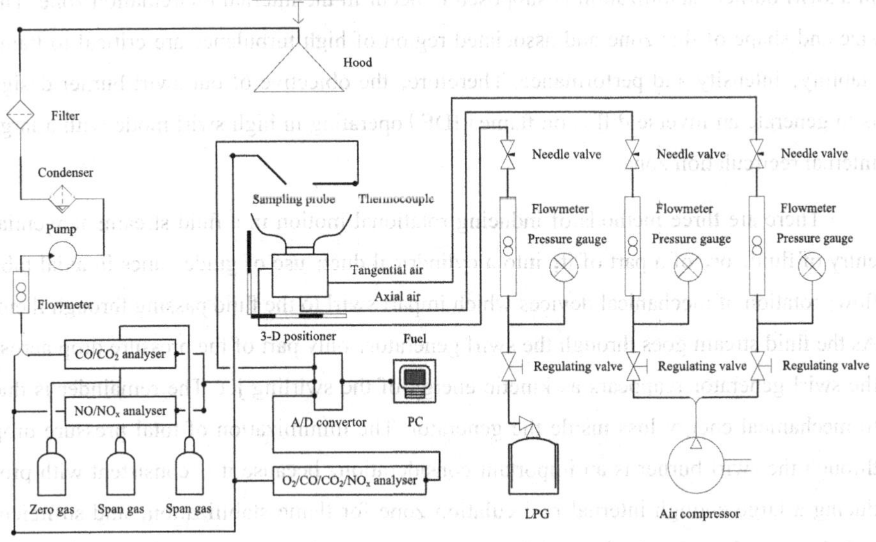

Figure 3.1 Experimental set-up for open flame studies.

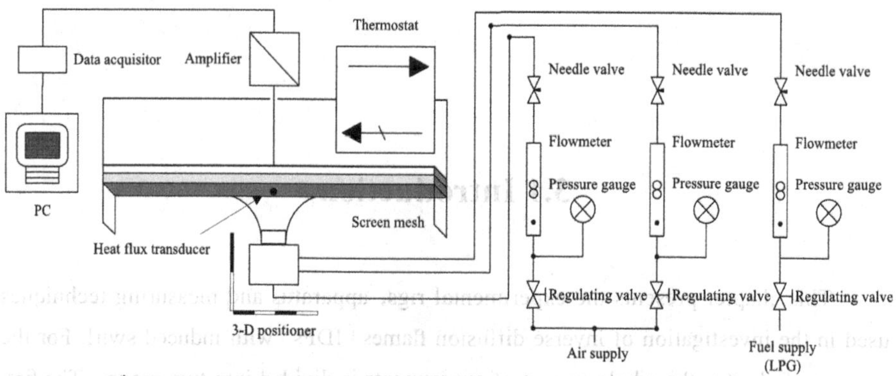

Figure 3.2 Experimental set-up for impinging flame studies.

3.2 Burner designs

The burner is the basic unit of any apparatus for flame study. Almost every research worker has his own design. A burner requires the burner geometry meet several restrictions, wherein operating conditions of the flame are influenced by the burner geometry. The burner must provide for stabilization of the flame. According to the literature review, in a swirl burner, stabilization is supposed to occur in the internal recirculation zone. The size and shape of this zone and associated region of high turbulence are critical to flame stability, intensity and performance. Therefore, the objective of our swirl burner design is to generate an inverse diffusion flame (IDF) operating in high swirl mode with a large internal recirculation zone.

There are three methods of inducing rotational motion in a fluid stream: tangential entry of fluid, or, of a part of it, into a cylindrical duct; use of guide vanes in axial tube flow; rotation of mechanical devices which impart swirl to the fluid passing through them. As the fluid stream goes through the swirl generator, only part of the pressure drop across the swirl generator reappears as kinetic energy of the swirling jet. The remainder is due to mechanical energy loss inside the generator. The minimization of total pressure drop through the swirl burner is an important consideration, because it is consistent with producing a large enough internal recirculation zone for flame stabilization, and sufficient turbulence and mixing in the initial region to promote fast rates, complete, and intense combustion. The swirl generation efficiency, defined as the ratio of the flux of kinetic energy of the swirling flow to the static pressure energy drop between the fluid inlet and the burner nozzle, shows substantial differences for various types of swirl generators. According to the review by Syred and Beer (1974), the generator of tangential entry including radial vanes usually gives high efficiency of 70%to 80%, with low pressure loss coefficients. As far as the method of tangential entry is concerned, usually at least two inlets are symmetrically spaced for an even flow structure in the swirl chamber. The inlet area ratio, A_t / A_e, appreciably affects the tangential and axial velocity profiles in the chamber, and the tangential nozzles should be positioned at some distance from the top of the cyclone, and recommended values are in the range a / L=0.08 – 0.5. Considerable reduction of loss coefficient occurs if the exit diameter is reduced compared to the outer diameter of the generator because the pressure loss coefficient from the inlets to the exhaust was shown to be mainly a function of outlet geometry and convergent exits in "Y" shape produce the lowest loss coefficients. The addition of a throat to the cyclone chamber enables similar kinetic energy of turbulence levels to be generated for much lower swirl number, which in turn means the total pressure drop through the swirl generator is much reduced. Typical range of the ratio D_e/D_0 used lies between 0.4 and 0.7. A divergent outlet appears to reduce the pressure loss coefficient, typically by 10%—20% and thus reduce substantially the swirl number necessary to cause the onset of the internal recirculation zone. With the divergent outlet, the reverse flow zone depends principally on both the degree of swirl and the angle of divergence of the outlet. Moreover, the reverse flow zone can be considerably enlarged in size by the divergent outlet and it was shown that a short divergent more than doubles, while a long divergent more than trebles the maximum diameter of the reverse flow zone.

Based on the findings and suggestions above, the final design of the swirl burner is completed. As shown in Figure 3.3, the swirl burner has a 12-mm-diameter central air port and twelve 2.4-mm-diameter outer fuel ports. Swirling motion is introduced into the central air stream by two tangential inlets into a cylindrical swirl chamber. There is also one axial inlet of air to the swirl chamber, which allows the degree of swirl to be changed by varying the ratio of axial to tangential flow rates. The axial air passes through a honeycomb flow straightener to produce a uniform axial stream. The tangential air is introduced through two tangential inlets to impart rotational motion, well upstream from the burner throat with A_t / A_e=6.25/36 and a / L=0.5. The large central air port, or the relatively large difference in diameters of the fuel and air port, is to ensure a large internal recirculation zone to form because it is found, during the development phase of this burner, that the size of the internal recirculation zone is dictated by the diameter of the air port. Each fuel port is at an angle of 45° to the air port, so that the fuel jets could strike the air jet and thus enhance the mixing between fuel and air. A short divergent exit is attached at right angles with the fuel ports with L_{div}/D_e=5/12, because a long one with L_{div}/D_e=1 is found to cause heating up of the burner by the flame. The burner head is made of brass and the burner body aluminum. Different burner heads can be installed onto the burner body which contains the swirl generator, i.e. the swirl chamber, thus facilitating the comparison of different flames at the same operating conditions.

D_e: Exit or air port diameter
D_o: Cyclone chamber diameter
L_{div}: Divergent exit length
L: Cyclone chamber length
A_t: Tangential inlet area
A_e: Exit area or air port area
a: Distance from the top

D_e=12 mm
D_e/D_o=1/2
L_{div}/D_e=5/12
L=48 mm
a/L=0.5
A_t/A_e=6.25/36

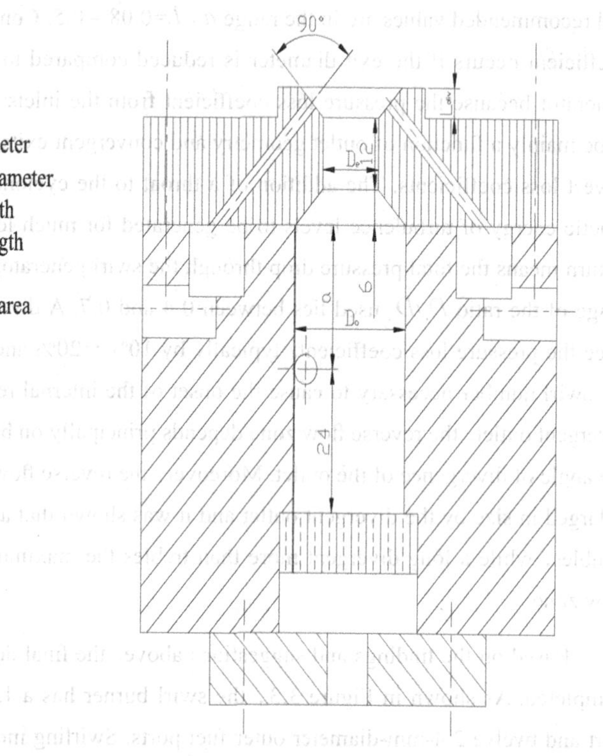

Figure 3.3 Construction and details of the swirl burner.

A non-swirling IDF burner with co-flowing fuel and air jets is also fabricated and shown in Figure 3.4. The diameters of both air/fuel ports and the center-to-center spacing between the fuel and air ports are the same to those of the swirl burner, so that direct comparison can be made between the swirling and non-swirling IDFs operating under the same conditions. For comparison of the swirling IDF and the swirling premixed flame, the swirl burner is modified by sealing the fuel ports by plasticine so as to produce a swirling premixed flame by using the central port ducting the premixed air/fuel mixture.

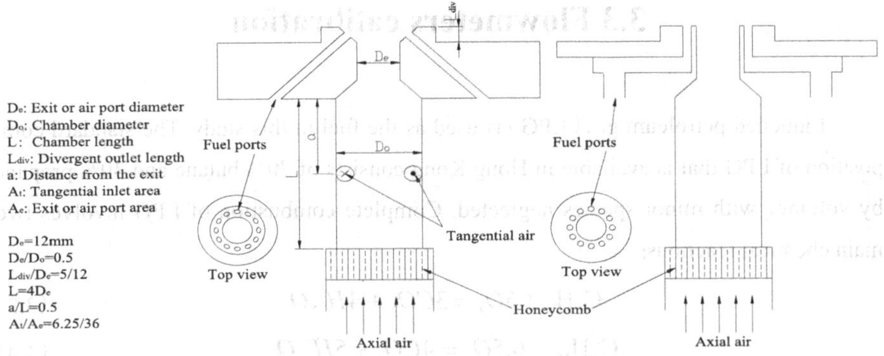

Figure 3.4 Schematic of the swirl and non-swirling burner.

The swirl number, whose definition is given in Equations 2.2, 2.3 and 2.4, should be calculated, if possible, from measured values of velocity and static pressure profiles. However, it is found that the swirl number can be satisfactorily calculated from input velocity distributions because once the steady flow has established in the burner, both the axial flux of angular momentum and the axial flux of linear momentum are conserved and do not vary as the cross-sectional area of the burner changes. Therefore, the swirl number can be calculated from the geometry of a swirl generator involving tangential and axial flows, with the pressure term omitted (Claypole and Syred 1981):

$$S \approx S' = \frac{G_\varphi}{G'_x R} = \frac{(D_0 - D_t) D_o}{2 D_t^2} \left(\frac{\dot{Q}_t}{\dot{Q}_t + \dot{Q}_a} \right)^2 = 9.12 \left(\frac{\dot{Q}_t}{\dot{Q}_t + \dot{Q}_a} \right)^2 \tag{3.1}$$

$$G_x' = \int_0^R \rho U^2 2\pi r dr \tag{3.2}$$

where $\dot{Q}_t$ and $\dot{Q}_a$ are the tangential and axial volumetric flow rates respectively, and D_0 and D_t are the chamber diameter and tangential inlet diameter respectively. With no axial inlet and only tangential inlets, the geometric swirl number is calculated to be the highest at 9.12, meaning that the degree of swirl can be changed from 0 to 9.12 by varying the relative magnitudes of the axial and tangential air flow rates on this particular swirl burner.

3.3 Flowmeters calibration

Liquefied petroleum gas (LPG) is used as the fuel in this study. The standard composition of LPG that is available in Hong Kong consists of 70% butane and 30% propane by volume, with minor species neglected. Complete combustion of LPG involves two main chemical reactions:

$$C_3H_8 + 5O_2 = 3CO_2 + 4H_2O \tag{3.3}$$

$$C_4H_{10} + 6.5O_2 = 4CO_2 + 5H_2O \tag{3.4}$$

So for stoichiometric combustion of 1 mole LPG, the amount of oxygen required is 6.05 moles. It means that 28.8 moles air is needed, the oxygen fraction of air being 21% by volume.

Theoretically, the stoichiometric volume ratio of air to fuel is 28.8, where oxygen comes from both the supply and the environment. Practically, the amount of oxygen from the environment is negligible when compared to that from the supply. Therefore, a concept of global or overall equivalence ratio, where only the supplied air is considered, has been used in most studies of non-premixed flames, which is also adopted in the this study:

$$\phi = \frac{x_{fuel} / x_{air}}{x_{fuel,stoich.} / x_{air,stoich.}} = \frac{28.8Q_{fuel}}{Q_{air}} \tag{3.5}$$

where the mole fractions of fuel and air are:

$$x_{fuel,stoich.} = \frac{1}{1+28.8} \tag{3.6}$$

$$x_{air,stoich.} = \frac{28.8}{1+28.8} \tag{3.7}$$

Both fuel and air can be treated as perfect gas at ambient conditions under which all the experiments in the present study are performed. The laboratory is ventilated and air-conditioned to a constant temperature at 21 ℃. The density of LPG is computed as:

$$\rho = \frac{P(44\times 30\% + 58\times 70)}{RT} = \frac{101325\times 53.8}{R\bullet 294} = 2.2Kg/m^3 \tag{3.8}$$

and the air density is 1.2 Kg/m^3. The Reynolds number of the air flow out of the burner head is:

$$Re = \frac{\rho V D_e}{\mu} = \frac{1.2\times 4Q_{air}}{1.8\times 10^{-5}\pi D_e} = 7073762Q_{air} \tag{3.9}$$

Equation 3.9 evaluates the Reynolds number in non-reacting conditions and the values of Reynolds number in reacting conditions decrease because the Reynolds number is inversely proportional to the kinetic viscosity and the value of the kinetic viscosity increases drastically in the combustion region.

Both fuel and air flow rates are stabilized by pressure gauges and metered by flowmeters. For determining the values of governing parameters of Re and Φ, flowmeters must be calibrated before use. The calibration of the fuel flowmeter (Matheson Type 605) with the pressure gauge adjusted to 1 kg/cm^2 is implemented by means of DryCal DC-2 (Bios International Corporation). DryCal DC-2 has three cells covering a flow range from 1 to 30 ml/min and only works on air flow. Air flow accuracy is 1 percent of averaged readings within a temperature range 0~55℃ and a humidity range 0~70%. Firstly, a relationship between graduations and flow rates of air is acquired, namely the calibration of the flowmeter by using air. Then, according to the variable area principle on which the flowmeter works: $\rho_{air}V_{air}^2=\rho_{fuel}V_{fuel}^2$, the relationship between graduations and flow rates of fuel can be derived, where the density ratio of air to fuel is 0.55. Another method of calibration of the fuel flowmeter is to measure mass flow rates of fuel by weighing the LPG tank at different graduations. Then volumetric flow rates can be obtained by dividing mass flow rates by the density of LPG. Computational results show that the data given by the two methods of calibration are in very good agreement.

DryCal DC-2 is incompetent for the calibration of the air flowmeters, Type 32121 and Type N044-40, owing to its small capacity. The following alternative method (BS 1042) is adopted which aims at reaching, at a 95% confidence level, an uncertainty on the flow rate smaller than ± 2%. The principle of the method is to determine the volumetric flow ratc by intcgration of thc vclocity distribution over the cross-sectional area of a pipe, which is illustrated in Figure 3.5.

The flowmeter is connected by a diverging entrance to a straight conduit to ensure that no separation of flow occurs. Flow is parallel to and symmetrical about the conduit axis and contains neither excessive turbulence nor swirl. The hatched ellipse in the figure designates the measuring cross-section which is perpendicular to the flow direction and is carved up into five concentric annuli of equal areas. Every intersection between five circles concentric with the conduit axis and two mutually perpendicular diameters are defined as the measuring points. A Pitot tube and a micro-manometer are used for velocity measurements.

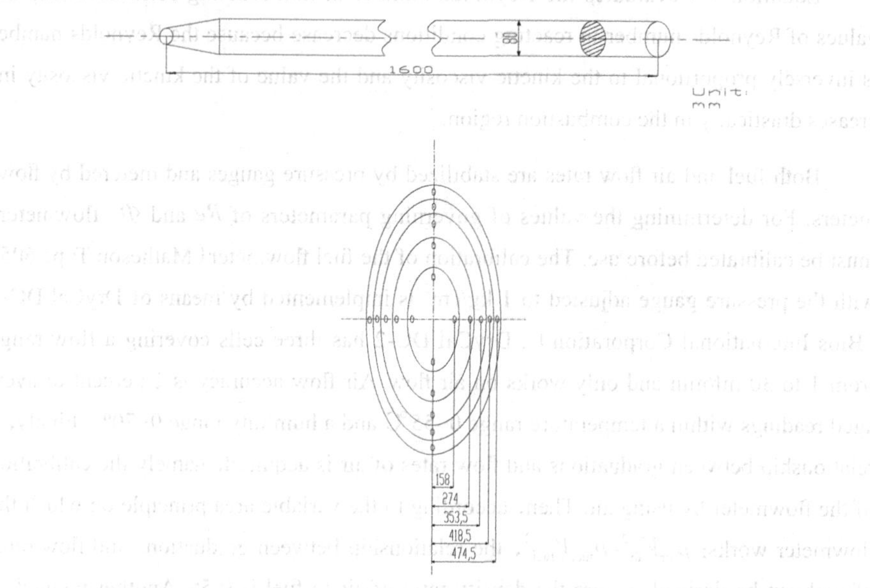

Figure 3.5 Pipe for calibration of the air flowmeters.

The calibrations of the air flowmeters for tangential supply and axial supply are implemented with the pressure gauges adjusted to 4 kg/cm^2 and 5 kg/cm^2, respectively. After the relationships between graduations and flowrates are obtained by use of the method mentioned above, confirmation is made to check the error caused by graphical integration. The lowest flow rate determined by the method is measured again by using DryCal DC-2. The error is found to be as small as 0.26%.

3.4 Temperature measurement

Flame temperature measurements provide important information on heat release as a result of chemical reactions in the flame. Temperatures can be easily measured by thermocouples. When two dissimilar conductors are connected electrically at two junctions maintained at different temperatures, a potential is developed which is proportional to the difference in temperature. This potential is a function of the materials chosen for the wires of the thermocouple.

To measure temperatures with high accuracy, thermocouples must be chosen, constructed and used with care. Type B thermocouple is most widely used in flame temperature measurements and is sometimes referred to by its nominal chemical composition. The positive thermoelement typically contains 29.60 ± 0.2% rhodium and the negative thermoelement usually contains 6.12 ± 0.02% rhodium. The 30-6 thermocouple can be used intermittently (for several hours) at temperatures up to 1790 ℃ and continuously (for several hundred hours) up to about 1700℃ with only small changes in calibration. The maximum temperature limit is governed primarily by the melting point of the Pt-6% rhodium thermoelement which is estimated to be 1820℃. The stability of the thermocouple at high temperatures depends primarily on the quality of the materials used for protecting and insulating the thermocouple. The thermocouple is the most reliable when used in a clean oxidizing atmosphere (air) but also has been used successfully in neutral atmospheres or vacuum. To facilitate the use of this thermocouple, after the thermocouple is calibrated by comparison with platinum resistance thermometers, a reference function and table are determined from measurements of the thermoelectric voltage at various thermometric fixed points over the range from 0 ℃ to 1790 ℃. In most applications the reference junction temperature does not need to be controlled or even known, as long as it is between 0℃ and 50℃.

Causes of errors between the true temperature and the temperature recorded by a thermocouple can be separated into perturbations by the probe and the errors involved in the particular technique employed. Perturbations by the probe can be classified as aerodynamic, thermal and chemical. To reduce this effect, wires should be selected having as small a diameter as possible, particularly when they are not subjected to mechanical strain. The aerodynamic effect is that the velocity-deficient wake behind the probe results in a local propagation of the flame, inducing a higher recorded temperature. This effect is proportional to the wire diameter. The thermal effect is that the probe as a heat sink is a function of the temperature difference between the gaseous stream and the thermocouple. This effect is not a problem since the difference is small for reliable thermocouples. The chemical effect is that the promotion of catalytic reactions on the probe surface results in spuriously high temperatures and hysteresis in the temperature profile. This effect can be made negligible by coating the probe with non-catalytic materials. There is another error, the movement or vibration of the probe which may offer a serious problem. This problem is best avoided by using heavier and shorter support wires. A thermocouple will register a temperature with radiation and conduction losses. These errors are involved in the technique and can be minimized with corrections. Conduction loss can be made negligible by aligning the supporting wires with a constant temperature surface. Radiation loss remains as an important source of error, and the loss can be completely eliminated by heating electrically the thermocouple to balance the radiation loss. Also, corrections can be making a balance between energy received from the gas and that lost by radiation.

The junction temperature of a thermocouple inserted into a gaseous stream is the resultant temperature of: heat transfer by convection between the thermocouple and the gas; heat transfer by radiation between the thermocouple, the gas and suspended particles; transfer of kinetic energy of the gas to thermal energy within the boundary layer of the junction; heat transfer by conduction along the wires when they pass through a temperature gradient which is not negligible. Heat transfer by convection, which we wish to augment, increases with the difference in temperatures of the gas and the thermocouple and also increases with the convection coefficient. To obtain a greater exchange of heat by convection, the arrangement of a thermocouple placed normal to the flow direction is preferable to the arrangement of a thermocouple placed along the flow direction. In addition, the convection coefficient increases with velocity and decreases as the diameter of the thermocouple is increased. Also, it increases considerably with turbulence. Heat transfer by radiation, which we wish to maintain as low as possible, can be reduced by using refractory sheaths. Polished metal surfaces have a small emissivity at low temperatures but the emissivity increases rapidly with temperature. Refractory sheaths (Al_2O_3) have an emissivity that decreases as the temperature increases and they are much less sensitive to chemical actions by the gas. Thus reducing radiation heat transfer can be achieved by surrounding the thermocouple with a system of Al_2O_3 shields. The error that the thermocouple measures the stagnation temperature is not significant since the energy exchange within the boundary layer is small. To reduce the error due to conduction, it is necessary to immerse the thermocouple wires in the flow with as great a length as possible of small diameter and with a low thermal conductivity.

For this study, uncoated 30-6 type B thermocouples are hand-made with care and employed for flame temperature measurements. The platinum-rhodium wires have a diameter of 0.25 mm and their joint bead is no more than 50 percent bigger than the wire diameter. Such sizes of the wires and hot junction are small enough to reduce the errors caused by perturbations of the thermocouple probe, and meanwhile, are big enough to keep the probe strong when subjected to any mechanical strain. The probe is curved into a "V" shape and has a length of less than 5 cm to avoid any excessive movement or vibration of the probe. During the course of experimenting, the probe is placed horizontally and thus normal to the flow direction to increase the convection heat transfer coefficient.

With a data acquisition module, Personal Daq/56 ™, which has built-in cold-junction compensation and converts measured thermoelectric voltages into temperatures, direct thermocouple measurements are made with spatial resolution of less than 0.05 mm^3 and temporal resolution of the order of 5 ms.

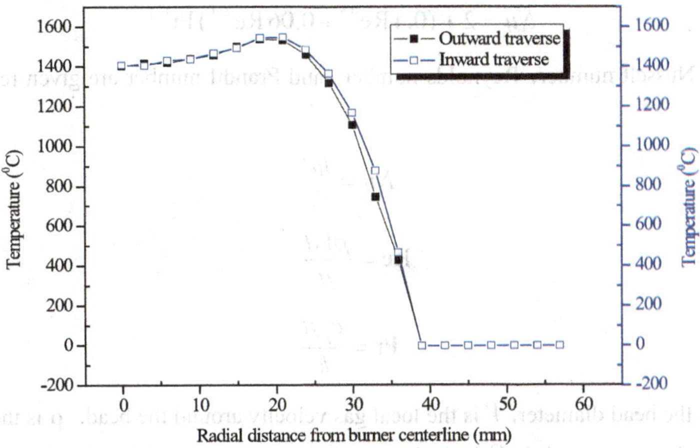

Figure 3.6 Temperature traverses for detection of the catalytic effect.

To check the catalytic effect which might have an influence on temperature data gained by uncoated thermocouples, several measurements are conducted by moving both the same thermocouple and different thermocouples to and fro along the radial direction through the flame. One pair of thermocouple traverses is shown in Figure 3.6. The difference in temperature only is found in the high temperature gradient region around the flame boundary. The reason might be the limit of spatial resolution of the thermocouple bead and the fluctuation in temperature caused by intense turbulence of the flame, rather than the catalytic effect. Therefore, it concludes that temperature profiles gained are successive and reproducible and the catalytic effect is not prominent for the flame under consideration.

Since radiation and conduction losses are important sources of error, all the averaged temperatures are corrected, based on the bare-bead thermocouple model proposed by Brohez *et al.* (2004).

For an uncoated spherical thermocouple bead in a hot gas stream, an energy balance in steady state yields:

$$h(T_g - T) = \varepsilon\sigma(T^4 - T_\infty^4) \tag{3.10}$$

where T_∞ is the temperature of surroundings, T is the thermocouple temperature, T_g is the true gas temperature, ε is the bare bead emissivity, σ is the Stefan-Boltzmann constant and h is the convective heat transfer coefficient.

For the purpose of estimating ε and h, the following correlations are used:

$$Nu = 2 + (0.4\mathrm{Re}^{0.5} + 0.06\mathrm{Re}^{2/3})\mathrm{Pr}^{0.4} \tag{3.11}$$

the Nusselt number, Reynolds number, and Prandtl number are given respectively by:

$$Nu = \frac{hd}{k} \tag{3.12}$$

$$\mathrm{Re} = \frac{\rho V d}{\mu} \tag{3.13}$$

$$\mathrm{Pr} = \frac{c_p \mu}{k} \tag{3.14}$$

d is the bead diameter, V is the local gas velocity around the bead, ρ is the gas density, μ is the gas viscosity, C_p is the heat capacity of the gas and k is the thermal conductivity of the gas.

The thermophysical properties of the air/fuel mixture are based on those of air due to the difficulty of estimating thermal and transport properties of multi-component mixtures. For air, we have the following relationships according to the book by Holman (2010):

$$\rho = PM / RT \tag{3.15}$$

$$\mu = 1.3554\times10^{-6} + 0.6738\times10^{-7}T - 3.808\times10^{-11}T^2 + 1.183\times10^{-14}T^3 \tag{3.16}$$

$$c_p = 1.05 - 3.736\times10^{-4}T + 9.34\times10^{-7}T^2 - 5.97\times10^{-10}T^3 + 1.293\times10^{-13}T^4 \tag{3.17}$$

$$k = -5.996\times10^{-4} + 1.015\times10^{-4}T - 4.56\times10^{-8}T^2 + 1.31\times10^{-11}T^3 \tag{3.18}$$

An iterative procedure is required to calculate the corrected temperature because the gas properties used to calculate the convection coefficient is dependent on the corrected temperature itself. The results show that heat is radiated and/or conducted away from the thermocouple, and the corrected temperature or the true temperature is always higher than the directly registered temperature. The higher the flame temperature is, the bigger the difference between the registered and true temperatures is. The maximum value in correction is 140℃ at the directly measured temperature of 1597℃ for the swirling IDF, indicating an error of 8.7% coherent in the technique.

3.5 Flame photography

Flame appearance or simply flame color is indicative of whether the flame is diffusional or premixed in nature. Diffusional combustion generates a bright, yellowish flame due to the high soot loading which gives off blackbody radiation. Premixed flame, however, is in the color of dim blue due to single wavelength radiation from various electron transitions in the excited molecules formed in the flame. Three techniques used to visualize flames photographically are direct, Schlieren, and shadowgraph photography. For this study, the method of direct photography with a digital CCD camera is employed. Because of the difficulty to record the sky blue flame region with any single exposure, the camera is operating in the videocorder mode, with 60 frames per second to record the luminous flame in a dark background. After recording, individual flame images are extracted from the videotape.

Flame length is an important parameter of a flame, which defines the high temperature zone and the time available for pollutant formation. The most commonly accepted definition of flame height is the vertical distance from the burner rim to the position where the fuel and oxidizer are in stoichiometric proportions (Mikofski *et al.* 2006). The suggested methods for flame length measurements include: visually inspecting the blue reaction zone; measuring the peak blue intensity recorded with a camera; visually inspecting the luminous yellow region in lightly sooting, non-smoking flames; using gas sampling; and locating the maximum temperature along the flame axis. For this study, three methods of determination of flame length have been tried out: highest temperature on the flame centerline, gas sampling of CO and visible flame length.

Flame length measurement can be the measurement of the highest flame temperature on the flame centerline, based on the fact that the flame temperature peaks at or near stoichiometric conditions. When the thermocouple registers the highest temperature on the flame centerline, the flame length is thus determined by the distance from the thermocouple bead to the burner rim. But, some measured data show a large deviation from the visible length of the flame measured by means of photography.

Flame length measurement can also be measuring the sharp reduction in CO concentration, based on the fact that intense combustion is associated with high CO concentration and CO is converted to CO_2 in the post flame region. Therefore, the sharp reduction in CO concentration occurs when the sampling probe is located at the flame tip. In an attempt to locate the sampling probe along the flame centerline in a position where the CO concentration decreases below 100 ppm, it is found that for very lean flames, the CO concentration sustains a long vertical distance in the post flame region. Therefore, the flame length acquired by this method is far longer than the visible flame length in fuel lean conditions.

The definition of luminous flame length is the vertical distance from the burner rim to the position where flame luminosity is no longer visible to the eye. Luminous flame length can be easily measured by means of direct flame photography. For this study, a digital CCD camera operating at 60 frames per second is used and the flame length is determined by averaging a dozen images of the visible flame operating under the same conditions.

3.6 In-flame gaseous species measurement

Composition of the gas within a flame provides information about the combustion process in the flame. The individual particles present in flame gases are commonly classified as molecules, free radicals, atoms and ions. Molecules refer to polyatomic species which are stable enough to be kept in pure form at ordinary temperatures and pressures, for example, H_2O, CO_2, CO, H_2, etc. Free radicals are molecular fragments — generally considered to be polyatomic—which are characterized by possession of one or more unpaired electrons, although they are electrically neutral. The unpaired electron means that gaseous free radicals are unstable or highly reactive and cannot be kept in pure form. Common examples are CH_3, OH, C_3H_7, CH_2, etc. A free radical consisting of just a single atom is usually referred to as simply an atom or an unstable atom to distinguish it from a stable atom such as is found in the rare gases in which all electrons are paired. The unstable atoms of interest in flames are those derived from the elements usually existing as diatomic gases, that is, H, O, N, Br etc. The final category refers to any species carrying a net charge, which is also unstable. A typical flame might consist of molecules, radicals and atoms, and ions in the proportions 1: 10^{-2}: 10^{-7}.

The determination of concentrations of stable molecules in a flame is usually carried out by two techniques: probe sampling followed by analysis or in situ analysis by spectroscopic means. In probe sampling, the sample is withdrawn, quenched and analyzed. Most of the problems associated with a flame application have been solved by the use of a very fine microprobe made of quartz and mass spectrometer or other micro-analytical techniques. Such a probe withdraws only a few micrograms of sample per second and does not disturb the flame visibly. The nozzle generating large pressure drop effectively quenches the flame reactions so that a reliable sample reaches the analytical instrument.

Sampling probes must produce a minimum of disturbance in the flame and allow rapid decompression and withdrawal of the sample to a cool region outside the flame. These can be met by a tapered quartz microprobe with a small orifice inlet. Tapers of 15° to 45° are satisfactory. The connecting tubing is to handle the inlet mass flow at a five – to tenfold pressure drop across the inlet orifice. The size of tubing is not critical since it has no disturbance to the flame. The orifice is chosen to be the smallest, yielding a mass flow of only a few micrograms per second.

Samples can either be taken in batches kept in suitable bottles, or introduced directly into an analytical instrument in a continuous flow arrangement. The pressure drop across the orifice must exceed a critical value so that the diffusion pump is isolated from the sample seen by the opening to the analytical instrument. The fraction of the sample taken by the mass spectrometer should be small ($\langle$ 0.01) to avoid biases. The pumping speed depends on the probe size and pressure in the sampling lines, but a small diffusion pump of a few liters per second at 25 μ pressure is usually adequate.

In the present study, the measurements of the composition of the gas within the flame are carried out by the probe sampling technique followed by analysis, or in situ analysis. Polyatomic gaseous species of O_2, CO_2, CO and NO_x, which are main parts of the flame gas, are considered and their concentrations are measured. A 15° to 45° tapered quartz probe with an inner diameter of 1 mm and an outer diameter of 2 mm is hand-made with care. The probe is inserted along the streamline direction from the hot side into the flame to produce a minimal disturbance and no visual disturbance is observed. The tapered orifice of the probe engenders a rapid pressure and temperature drop in the sample gas flow to quench chemical reactions and "freeze" the sample composition. The quartz probe is 20 cm long and is connected to a stainless steel tube with an inner diameter of 1 cm and length of 1 m to handle the rapid cooling of the sample. The gas sample effluent from this 1-m long tube cools down below 60℃. The sampling rate is controlled by the pump in the analyzer. In consideration of reducing the fraction of the sample taken by the analyzer to avoid biases, two runs of measurements are conducted. One run is performed at the pumping speed of 5 to 6 liters per minute by using an exhaust gas analyzer: Anapol EU-5000 (Anapol Instruments Engineering Incorporation) for O_2 (Electrochemical cell) and CO (NDIR) measurements. The other run is for NO and NO_x measurements by use of a chemiluminescence NO/NO_x analyzer (California Instruments Corporation, Model 400 HCLD, CLA) and for CO_2 measurements with a CO_2 analyzer (California Instruments Corporation, Model 300, NDIR), where the sampling rate is controlled by a membrane pump working at 3 liters per minute. The output signals of each analyzer are processed by amplifiers, digitized by an analog/digital converter, converted into the final results and recorded by a computer. All the measurements start only after an equilibrium state is reached within the flame-analyzer system. For Anapol EU-5000, the waiting time between two successive measurements is around 60 seconds, meaning that after the sampling probe has been moved to a new measuring point, the steady state in the flame-analyzer system needs 60 seconds to establish. For each measuring point in the flame, 60 data are registered and their averaged value is reported. For the NO/NO_x analyzer, the waiting time is around 20 seconds. In addition, zero and span calibrations are performed before and after each measurement.

3.7 Flue gas measurement

Overall pollutant emission characteristics of a flame can be acquired by flue gas measurements, with the results expressed in terms of emission index. Emission index is generally used for comparison of flame emissions, especially for NO_x emissions. For calculation of NO_x emission index, simultaneous concentration measurements of NO_x and CO_2 are required. For this study, two methods of flue gas measurement: hood and probe sampling, are used and compared. Besides NO_x, pollutants including CO, NO and NO_2 are also considered. Assuming that essentially all of the carbon from the fuel is oxidized to CO_2, the emission index can be calculated from:

$$EI_J = \frac{3.7\{J\}MW_J 1000}{\{CO_2\}MW_{C_{3.7}H_{9.4}}} \tag{3.10}$$

Where the brackets indicate mole fraction, MW is molecular weight, and J symbolizes the pollutant species.

As for the probe sampling method, both stainless steel and quartz probes are used, and they are air-cooled instead of water-cooled. The probe is translated both axially and radially. Therefore both axial and radial profiles of concentration are obtained. From radical profiles, it is seen that there exists a central core region in which the calculated emission index is nearly constant with only a little drop away from the burner axis. Outside this core which is approximately 12 mm in radius, species are diluted by ambient air to a larger extent, leading the emission index to fall down sharply. Samples are extracted at locations 50 mm to 150 mm above the visible flame tip. In this range of height, no difference in the concentration is found by means of either stainless steel probe or quartz probe. Accordingly, the emission index is calculated to be constant.

With regard to the hood method, two different hoods, a cylinder and a cone, are adopted. The diameter of the cylindrical hood is twice that of the conical hood. Under the same experimental condition, the emission index obtained by the use of the conical hood is found to be larger than that by using the cylindrical hood. The reason is that the bigger volume of the cylindrical hood allows more air to mix with the species, leading to a higher level of uneven dilution. Compared to the cylindrical hood, the conical hood can generates an even flow inside the core. The comparison of the probe sampling and hood methods indicates that the RMS value of emission index obtained by probe sampling is larger. Therefore, the conclusion is made that the hood method is better and thus chosen for future use.

The hood method proposed by Butcher *et al.* (1984) and validated by G. Ballard *et al.* (1999) is used. A stainless steel hood of 15 cm diameter and 15 cm depth collects exhaust products in the post-combustion region and samples at a vertical distance of 120 mm above the visible flame tip are measured. The same air-cooling 1m long stainless steel tube is used. Water vapor is condensed and removed from the sample gas before it enters the chemiluminescent NO/NO_x analyzer (California Instruments Corporation, Model 400 HCLD) and $CO/CO_2/SO_2$ analyzer (California Instruments Corporation, Model 300). Zero and span calibrations are performed before and after each measurement. The emission indices of $NO/NO_2/NO_x$ and CO are calculated from the measured data. NO_2 concentration is obtained by subtracting sequential measurements of NO_x and NO concentrations and the molecular weight of NO_2 is used for NO_x emission index calculation.

3.8 Heat transfer measurement

Flame impingement heat transfer has been widely used in industrial and domestic heating applications for its excellent thermal performance. Developments in thermopile construction have produced small sensors that can be used to make localized measurements of heat flux. For this study, a heat flux sensor (Vatell Corporation, Model HFM-6D/H) has been adopted to explore and understand the local heat flux from impinging flame jets.

Local heat flux measurements are performed using a rig consisting of a thermostat, a heat flux transducer and an impingement copper plate, as shown in Figure 3.2. The burner is mounted on a 3-D positioner which can move the burner to any position in space. For minimization of the disturbance from the surrounding air flow, a screen mesh is used to enclose the flame. The flame impinges vertically onto the target surface which is a water-cooled copper plate with a surface area of 600 × 600 mm^2 and a thickness of 12 mm. The copper plate is evenly covered and cooled by a water jacket and the temperature of the cooling water is kept constantly at 38℃ by a thermostat (Julabo Labortechnik GmbH, Model fc1600s) to avoid condensation. The top plate covering the cooling water jacket is made of plexi-glass to enable the water flow visible. The heat flux transducer is attached to the center of the copper plate to measure the local heat flux from the flame to the target plate. It is a small ceramic heat flux transducer with an effective sensing area of 2 × 2 mm^2 and negligible thickness and its installation has been based on the manufacturer's specifications with a ± 0.01 mm dimensional tolerance to ensure that the sensor and plate are in the same alignment and the sensor surface is flush with the front side of the plate facing the flame. The standard coating for the face of the sensor is zynolyte, which is a high temperature black coating with a 0.94 emissivity. The coating allows both convection and radiation heat transfer to be sensed by the transducer. The output signals of the transducer are processed by a voltage amplifier (Vatell Corporation, Model AMP-6), digitized by the analog/digital converter (Personal Daq/56 ™), recorded by the computer and converted into the final results according to calibration data of the transducer provided by its manufacturer.

Water-boiling test is usually used by researchers because it gives a general idea of the heat transfer behavior of an impinging flame. For this study, a volume of 500 ml tap or pure water was poured into a cylindrical pot. The flame was positioned beneath the pot and the separation distance was varied by an XYZ-cage. A thermometer was suspended vertically over the pot with its sensing bead immersed in the water at a position with a certain radial distance off the geometric center of the pot. The temperature of the tap water was below 30 ℃ and the reading of the thermometer ascends steadily when the water is heated from 40℃ to 80℃ . The period of time for this steady warming up was registered by a digital calculagraph.

3.9 Error analysis

Measurement is the estimation of the magnitude of some attribute of an object, such as its length or weight, relative to a unit of measurement. A measurement is understood to have three parts: first, the measurement itself, second, the margin of error, and third, the confidence level — that is, the probability that the actual property of the physical object is within the margin of error.

The result of a measurement, i.e. an estimate of the value of the measurand is determined usually under repeatability conditions, where variations are caused by influence quantities. An error has two components, random and systematic. Random error causes variations in repeated observations of the measurand and can be reduced by increasing the number of observations. Systematic error derives from influence quantities, too, and only if their effect can be identified and quantified, a correction can be made.

The uncertainty in the result of a measurement reflects the lack of knowledge of the value of the measurand, and in practice, there are many sources of uncertainty. Uncertainty components are often grouped into categories A and B, which are two different ways of evaluating uncertainty components. Components A and B are quantified by variance and standard deviation. For type A evaluation, the estimated variance μ^2 is calculated from repeated observations and is the familiar statistically estimated variance s^2. The estimated standard deviation is the positive square root of μ^2, μ=s and is called Type A standard uncertainty. For type B evaluation, the estimated variance is obtained using available knowledge. When a result is obtained from a number of other quantities, its combined standard uncertainty is equal to the positive square root of the combined variance obtained from all variance and covariance, using the law of propagation of uncertainty.

A mathematical model of the measurement is usually used to evaluate the uncertainty. Because the mathematical model may be incomplete, all relevant quantities should be varied to the fullest extent and the model should be revised when the observed data demonstrate it is incomplete. In most cases, a result is determined from a number of other quantities, through a functional relationship:

$$Y=f(X_1, X_2, \cdots, X_n) \tag{3.11}$$

The input quantities may themselves be viewed as measurands and depend on other quantities. If f does not model the measurement to the required accuracy, additional input quantities must be included. The arithmetic mean obtained from n independent repeated observations is the input estimate of X_i:

$$\overline{X}_i=\frac{1}{n}\sum_{k=1}^{n}x_k \tag{3.12}$$

The input estimates which are not evaluated from repeated observations must be obtained by other methods. The estimate of variance is:

$$s^2=\frac{1}{n(n-1)}\sum_{k=1}^{n}(x_k-\overline{X}_i)^2 \tag{3.13}$$

Its positive square root is the standard uncertainty for X_i. The degrees of freedom are n-1 if there are n independent observations. For an input quantity that is not obtained from repeated observations, its variance and standard uncertainty are evaluated by scientific judgment based on available information, and a proper type B evaluation calls for insight based on experience and general knowledge. An uncertainty component that is included as a type B evaluation is an independent component of uncertainty in the calculation of the combined standard uncertainty of the measurement result. Y is estimated as:

$$y=f(\overline{X}_1,\overline{X}_2,\ldots,\overline{X}_N) \tag{3.14}$$

Its standard uncertainty or the combined standard uncertainty is:

$$u(Y)=\sqrt{\sum_{i=1}^{N}\left(\frac{\partial f}{\partial X_i}\right)^2 u^2(X_i)}\ . \tag{3.15}$$

Where $\frac{\partial f}{\partial X_i}$ is evaluated at $X_i=\overline{X}_i$. This is the law of propagation of uncertainty.

When the input quantities are correlated, the expression for the combined variance is:

$$u^2(Y)=\sum_{i=1}^{N}\sum_{j=1}^{N}\frac{\partial f}{\partial X_i}\frac{\partial f}{\partial X_j}u(X_i,X_j) \tag{3.16}$$

Where $u(X_i,X_j)=u(X_j,X_i)$ is the estimated covariance. The degree of correlation is characterized by the estimated correlation coefficient:

$$r(X_i,X_j)=\frac{u(X_i,X_j)}{u(X_i)u(X_j)} \tag{3.17}$$

Correlations between input quantities cannot be ignored if present and significant.

When reporting the result of a measurement and its uncertainty, one should: describe clearly the method used to calculate the measurement result and its uncertainty from the experimental observations or other input data; list all uncertainty components and describe how they are evaluated; present the data analysis in such a way that each important step can be readily followed and the calculation of the result can be repeated if necessary; give all corrections and constants used and their sources.

3.9.1 Uncertainty in Φ

The model of overall equivalence ratio is:

$$\phi=\frac{28.8Q_{fuel}}{Q_{air}} \tag{3.18}$$

Where Q_{fuel} and Q_{air} are metered from flowmeters Type 605 and Type 32121 in most cases. Type 605 is calibrated by DryCal DC-2 at 21℃ and 1 kg/cm^2. The air flow accuracy of DryCal DC-2 is 1 percent of averaged readings within a temperature range 0~55 ℃ and a humidity range 0~70%. At 101.3 kPa and 21℃, the correction factors for gases other than air are 1.42 for butane, 1.23 for propane and 1.35 for the presently used LPG. So for the calculated flow rates of LPG, an accuracy of ± 5% is estimated from the manufacture's specification. Type 32121 is calibrated by a method from British standard which generates an uncertainty in the flow rate smaller than ± 2% with a 95% confidence level. Therefore, the quoted uncertainties of Q_{fuel} and Q_{air} are ± 5% and ± 2%, respectively, of the nominal value with a 95 confidence level. Assume a normal distribution is used to calculate the quoted uncertainty and let all relevant quantities vary to the fullest practicable extent, we can recover the standard uncertainty of Q_{fuel} and Q_{air} . Two partial derivatives are:

$$\frac{\partial\phi}{\partial Q_{fuel}}=\frac{28.8}{Q_{air}} \tag{3.19}$$

$$\frac{\partial\phi}{\partial Q_{air}}=-\frac{28.8Q_{fuel}}{Q_{air}^2} \tag{3.20}$$

The two flowmeters are calibrated separately and when estimating the degree of correlation between input quantities arising from the effects of common influences, such as ambient temperature, barometric pressure, and humidity, such influences have negligible interdependence and the affected input quantities can be assumed to be uncorrelated. By using the law of propagation of uncertainty, the combined standard uncertainty of Φ is:

$$u(\phi)=[(\frac{\partial\phi}{\partial Q_{fuel}})^2 u(Q_{fuel})^2+(\frac{\partial\phi}{\partial Q_{air}})^2 u(Q_{air})^2]^{1/2}=\sqrt{0.6261\frac{Q_{fuel}^{\ 2}}{Q_{air}^{\ 2}}} \tag{3.21}$$

By putting the volumetric flow rates of air and fuel into Equation 3.21, the uncertainty can be obtained. With a 95 confidence level, the maximum uncertainty in Φ is 5%.

3.9.2 Uncertainty in *Re*

The model of Reynolds number is:

$$Re=\frac{\rho VD}{\mu} \tag{3.22}$$

Where the characteristic length is the diameter of the central port and the characteristic velocity is the mean velocity of air flow through the central port:

$$V=\frac{Q_{air}}{A}=8842Q_{air} \tag{3.23}$$

Because there is no combustion before fuel and air mix, thus we can assume the preheating of air flow in the central port is small, and the density and viscosity of air are constant. Therefore:

$$Re=7073762Q_{air} \tag{3.24}$$

The quoted uncertainty of Q_{air} is ± 2% of the nominal value with a 95 confidence level. One partial derivative is:

$$\frac{\partial Re}{\partial Q_{air}}=7073762 \tag{3.25}$$

The combined standard uncertainty is:

$$u(Re)=[(\frac{\partial Re}{\partial Q_{air}})^2 u(Q_{air})^2]^{1/2}=72181Q_{air} \tag{3.25}$$

By using a 95 confidence level, the uncertainty in *Re* is 2%.

3.9.3 Uncertainty in *T*

The model of temperature simply is:

$$T=T \tag{3.26}$$

Because temperature is measured by thermocouples, based on standard tables which show the relationship between the output voltage produced by thermocouples and any given temperature. Systematic errors must be identified and corrected. Detailed analysis of systematic errors has been done in Subsection 3.4 and the results show that the radiation and conduction loss is significant in size relative to the required accuracy of the measurement. The radiation and conduction loss has been quantified and correction is made to compensate its effect in Subsection 3.4. The estimate of T is the arithmetic mean of 50 independent observations obtained under the same measurement conditions, and the standard uncertainty is:

$$T = \sqrt{\frac{1}{2450}\sum_{k=1}^{50}(t_k - \bar{T})^2} \tag{3.27}$$

With the data from a single measurement, the value given by Equation 3.27 indicates the measurement repeatability. With the data from a set of individual measurements, Equation 3.27 gives the measurement reproducibility. The uncertainty in temperature is the combined standard uncertainty that considers the uncertainties caused respectively by the repeatability, reproducibility and systematic errors.

By using a 95 confidence level, the maximum and minimum uncertainties in temperature are 22% and 0.04%, respectively.

3.9.4 Uncertainty in flame length

The model of flame length is:

$$L = L \tag{3.28}$$

Because the flame length is simply obtained by comparing the visible flame with a standard length meter, the estimate of each flame length is the arithmetic mean of a dozen observations obtained under the same measurement conditions, and the standard uncertainty is:

$$L = \sqrt{\frac{1}{132}\sum_{k=1}^{12}(l_k - \bar{L})^2} \tag{3.29}$$

By using a 95 confidence level, the uncertainty in L is 10%.

3.9.5 Uncertainty in concentration

The model of O_2, CO, CO_2 and NO_x concentrations are the same:

$$C = C \tag{3.30}$$

For each test both zero and span gases have been used to calibrate the equipment to eradicate systematic errors. Hence, the uncertainty in concentration is the combined standard uncertainty that considers the uncertainties caused by the repeatability and reproducibility of the measurement. By using a 95% confidence level, the maximum and minimum uncertainties in O_2 concentration are 9.3% and 0%, respectively. The maximum and minimum uncertainties in CO concentration are 5.9% and 0%, respectively. The maximum and minimum uncertainties in CO_2 concentration are 4.8% and 0%, respectively. The maximum and minimum uncertainties in NO_x concentration are 7.3% and 0% respectively.

CHAPTER 4 FLAME CHARACTERISTICS AND TEMPERATURE

This chapter presents the results of the experimental investigation on the flame temperature field. The presentation includes the investigation of the flame appearance such as flame color, flame length and a qualitative analysis of the flame structure. Flame photographs and schematic diagrams are used as an aid to investigate the effect of governing parameters of the Reynolds number and the overall equivalence ratio on the flame temperature. Further investigations concerning the in-flame stable gaseous species and the overall pollutant emissions will be presented in Chapter 5.

4.1 Flame appearance

The swirl burner is designed to generate an IDF operating in high swirl mode with a large internal recirculation zone. Therefore, the experimental tests are carried out without the use of the axial air supply, and by only using the tangential air supply, the highest degree of swirl is produced and the swirl number is calculated to be 9.12. It is observed that when either the air flow rate or the fuel flow rate is varied, the visual features of the swirling IDF changes accordingly.

Table 4.1 Operational conditions.

S'	*Tangential air flow rate (32121) (10-4m-3/s)*	*Tangential air flowmeter (32121) reading*	*Axial air flow rate (N044-40) (10-4m-3/s)*	*Axial air flowmeter (N044-40) reading*	*Fuel flow rate (605) (10-5m-3/s)*	*Fuel flowmeter (605) reading*	*Re*	*Φ*
9.12	1.41	4	0	/	0.74	3	1000	1.5
9.12	2.83	9	0	/	1.47	6	2000	1.5
9.12	4.24	13	0	/	2.21	9	3000	1.5
9.12	5.65	18	0	/	2.95	12	4000	1.5
9.12	7.07	23	0	/	3.68	15	5000	1.5

续 表

S′	Tangential air flow rate (32121) (10-4m-3/s)	Tangential air flowmeter (32121) reading	Axial air flow rate (N044-40) (10-4m-3/s)	Axial air flowmeter (N044-40) reading	Fuel flow rate (605) (10-5m-3/s)	Fuel flowmeter (605) reading	Re	Φ
9.12	8.48	28	0	/	4.42	18	6000	1.5
9.12	9.90	33	0	/	5.15	20	7000	1.5
9.12	11.31	37	0	/	5.89	23	8000	1.5
9.12	12.72	42	0	/	6.63	26	9000	1.5
9.12	14.14	47	0	/	7.36	29	10000	1.5
9.12	15.55	51	0	/	8.10	32	11000	1.5
9.12	16.96	56	0	/	8.84	34	12000	1.5
9.12	18.38	61	0	/	9.57	37	13000	1.5
9.12	19.79	66	0	/	10.31	40	14000	1.5
9.12	21.21	70	0	/	11.04	43	15000	1.5
9.12	22.62	75	0	/	11.78	46	16000	1.5
9.12	24.03	80	0	/	12.52	48	17000	1.5
9.12	25.45	84	0	/	13.25	51	18000	1.5
9.12	26.86	89	0	/	13.99	54	19000	1.5
9.12	28.27	94	0	/	12.76	56	20000	1.5
9.12	11.31	37	0	/	3.93	16	8000	1.0
9.12	11.31	37	0	/	4.71	19	8000	1.2
9.12	11.31	37	0	/	5.49	22	8000	1.4
9.12	11.31	37	0	/	6.28	25	8000	1.6
9.12	11.31	37	0	/	7.07	28	8000	1.8
9.12	11.31	37	0	/	7.85	31	8000	2.0
9.12	8.48	28	0	/	5.89	23	6000	2.0

续 表

S′	*Tangential air flow rate (32121) (10-4m-3/s)*	*Tangential air flowmeter (32121) reading*	*Axial air flow rate (N044-40) (10-4m-3/s)*	*Axial air flowmeter (N044-40) reading*	*Fuel flow rate (605) (10-5m-3/s)*	*Fuel flowmeter (605) reading*	*Re*	*Φ*
9.12	1.41	4	0	/	0.25	1	1000	0.5
9.12	2.83	9	0	/	0.59	3	2000	0.6
9.12	4.24	13	0	/	1.03	4	3000	0.7
9.12	5.65	18	0	/	1.57	6	4000	0.8
9.12	7.07	23	0	/	2.21	9	5000	0.9
9.12	8.48	28	0	/	2.95	12	6000	1.0
9.12	9.90	33	0	/	3.44	14	7000	1.0
9.12	12.72	42	0	/	4.42	18	9000	1.0
9.12	14.14	47	0	/	5.40	21	10000	1.1
9.12	15.55	51	0	/	5.94	23	11000	1.1
9.12	16.96	56	0	/	6.48	25	12000	1.1
9.12	18.38	61	0	/	7.66	30	13000	1.2
9.12	19.79	66	0	/	8.25	32	14000	1.2
9.12	21.21	70	0	/	8.84	34	15000	1.2
9.12	22.62	75	0	/	9.42	37	16000	1.2
9.12	24.03	80	0	/	10.85	42	17000	1.3
9.12	25.45	84	0	/	11.49	44	18000	1.3
9.12	26.86	89	0	/	12.13	47	19000	1.3
9.12	28.27	94	0	/	12.76	49	20000	1.3
2.1	54.27	16	5.89	39	5.49	22	8000	1.4
2.2	55.55	17	5.75	38	5.49	22	8000	1.4
2.3	56.79	18	5.63	37	5.49	22	8000	1.4

续 表

S'	Tangential air flow rate (32121) (10-4m-3/s)	Tangential air flowmeter (32121) reading	Axial air flow rate (N044-40) (10-4m-3/s)	Axial air flowmeter (N044-40) reading	Fuel flow rate (605) (10-5m-3/s)	Fuel flowmeter (605) reading	Re	Φ
2.4	58.02	19	5.51	36	5.49	22	8000	1.4
9.12	2.83	9	0	/	0.19	1	2000	0.2
9.12	2.83	9	0	/	0.49	2	2000	0.5
9.12	5.65	18	0	/	0.68	3	4000	0.3
9.12	5.65	18	0	/	0.94	4	4000	0.5
9.12	8.48	28	0	/	1.98	8	6000	0.7
9.12	8.48	28	0	/	2.24	9	6000	0.8
9.12	11.31	37	0	/	3.49	13	8000	0.8
9.12	11.31	37	0	/	3.53	14	8000	0.9
9.12	14.14	47	0	/	4.42	18	10000	0.9
9.12	14.14	47	0	/	4.91	19	10000	1.0
9.12	16.96	56	0	/	5.69	22	12000	0.9
9.12	16.96	56	0	/	5.89	23	12000	1.0
9.12	19.79	66	0	/	6.87	27	14000	1.0
9.12	19.79	66	0	/	7.55	30	14000	1.1

4.1.1 Varying *Re* at fixed *Φ*

For the simplification of observation of the flame, the Reynolds number (*Re*) is increased progressively from zero to the maximum value for a stable flame to establish, and meanwhile the overall equivalence ratio (*Φ*) is kept at 1.5. This means increasing the volumetric flow rates of both fuel and air and simultaneously keeping their ratio unchanged. The resultant flame images are shown in Figure 4.1.

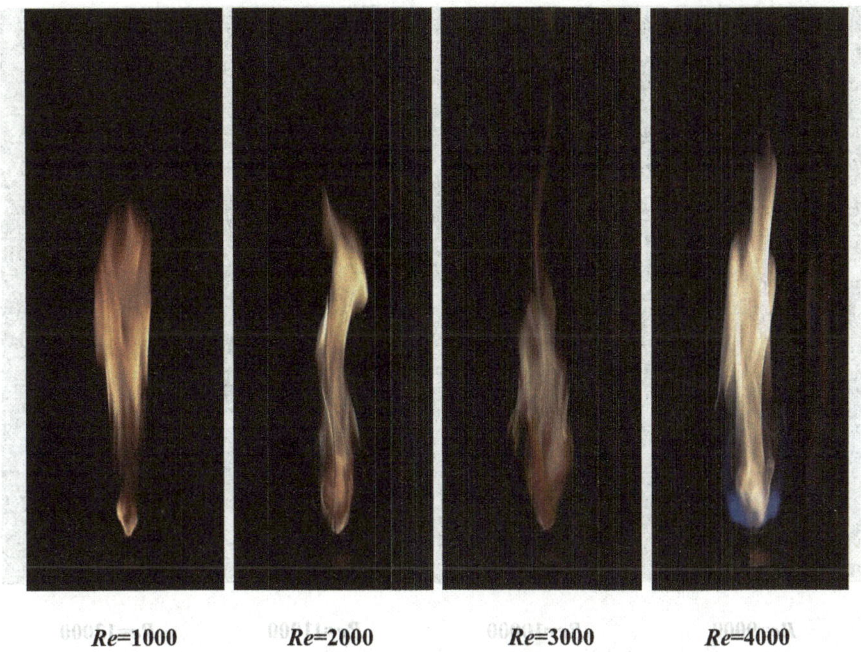

Re=1000 *Re*=2000 *Re*=3000 *Re*=4000

Re=5000 *Re*=6000 *Re*=7000 *Re*=8000

Re=9000 *Re*=10000 *Re*=11000 *Re*=12000

Re=13000 *Re*=14000 *Re*=15000 *Re*=16000

***Re*=17000 *Re*=18000 *Re*=19000 *Re*=20000**

Figure 4.1 Flame images at different *Re* at fixed *Φ*=1.5

The flame images shown in Figure 4.1 are extracted from the videotape recorded by the digital camera and the figure illustrates that at low values of $Re<4000$, the swirling flame consists of two parts: a peach-shaped flame root and a long flame tail. Both parts are in the colour of yellow, which is indicative of the occurrence of a diffusional combustion condition. An internal recirculation zone (IRZ) is formed in the centre of the flame, specifically in the centre of the flame root. The yellow colour makes the IRZ easily recognizable to the naked eyes. The IRZ is observed to occupy a large proportion of the main flow field (flame root) due to a large portion of the initial flow being recirculated, and two vortexes reside symmetrically to the burner nozzle axis due to the rotational motion of the fluid. As the value of Re increases, the flame root augments in size and the flame tail diminishes in both size and length. As Re increases beyond 6000, the long flame tail totally disappears and an entirely blue flame is formed, which is mainly premixed in nature because of more intensive mixing of the supplied fuel and air and hence more complete combustion. The blue colour makes the IRZ invisible to the eyes so that a flow visualization technique is of necessity for the purpose of observing the flame visually . The flow visualization is achieved by inserting a thin wooden rod into the flame to dye the fluid colour to yellow. The small yellow blaze introduced by combusting the rod follows the mainstream and acts as streaklines. Thus, the local flow fields are visible to the naked eyes. By placing the rod in different positions in the flame and combining the separate local flow fields, the complete flow field or the overall flame structure could be acquired. When the rod is inserted at a vertical distance of 15 mm from the burner rim, the local flow fields are shown by the photographs proceeding from left to right in Figure 4.2（a）. When the rod is close to the boundary of the flame, the yellow blaze develops upwards with its tip deflected towards the centerline of the flame. As the rod is further moved inwards, the blaze diverts its direction to the centerline of the flame and then develops downwards, such that a loop is formed. When the rod is at the centerline of the flame, the blaze directly diverts its direction downwards and then radially upwards so that a loop is also formed. Figure 4.2（b）illustrates another traverse of the rod at a vertical distance of 30 mm from the burner rim. Similarly, the yellow blaze first develops upwards with its tip deflected towards the centerline of the flame when the rod is close to the boundary of the flame. When the rod is further inward, the portion that is deflected is longer and when the rod is at the centerline of the flame, the blaze reverses its direction to develop downwards. A vertical traverse of the rod along the centerline of the flame is shown in Figure 4.2（c）, and it is seen that when the rod is close to the burner rim, the yellow blaze directly develops downwards; when the rod is at a vertical distance of around 60 mm from the burner rim, the separation of the blaze occurs, one part developing upwards and the other part developing downwards; when the rod is further away from the burner rim, the blaze develops directly upwards.

（a）

(b)

(c)

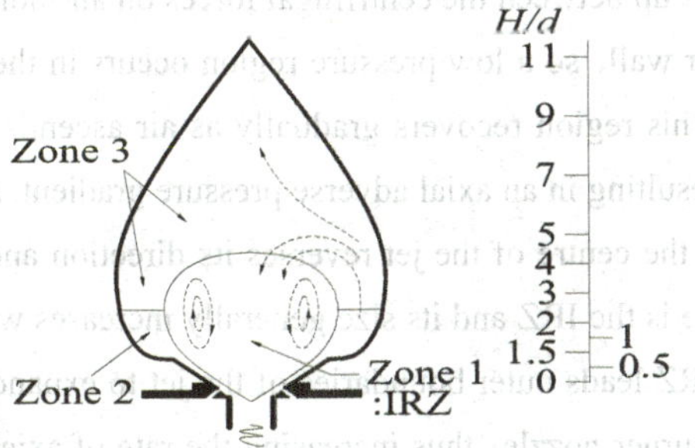

(d)

Figure 4.2 Images of flow visualization and structure of open swirling IDF at *Re*=8000, *Φ*=1.4 and *S'*=9.12.

The combination of the separate local flow fields in Figure 4.2 (a), (b) and (c) identifies the complete flow field inside the flame, as shown in Figure 4.2 (d), where dashed lines roughly sketch out the paths of fluid particles. It can be seen that fluid particles come out of the divergent burner outlet mainly through an annular region, and then go upwards along the boundary of the flame. Then a large portion of the fluid particles in the boundary are diverted towards the centerline of the flame and recirculate towards the burner exit to form a reverse flow. This flow structure is similar to that reported by Ji and Gore (2002), who studied the mean flow structure in strongly swirling flows. They stated that the reverse flow boundary is shaped like an ellipse, with the mean gas flow above the boundary going upstream and the mean gas flow below these lines going downstream. Based on the analysis of the complete flow field, the peach-shaped flame can be characterized into three distinctive zones: Zones 1, 2 and 3. Zone 1 is the reverse flow or IRZ occurring in the centre of the flame. The IRZ occupies a large proportion of the volume of the flame and two vortex eyes are seen to be located symmetrically to the burner nozzle axis. Since the flame is axisymmetrical, the two vortex eyes indicative of peak values of vorticity actually represent a 3-D annulus. Zone 2 is the annular region or the flame boundary in the lower section, being in contact with ambient air on the outer side and with Zone 1 on the inner side. It is always navy-blue in colour, indicating that premixed combustion is occurring in this zone. Zone 2 is the mixing zone in which the swirling air jet coming out from the air port mixes with the fuel coming out from the fuel jets. Zone 3 is the flame boundary in the upper section, most part of which has the same colour as Zone 1.

The IRZ is formed in the centre of the flame at $Re>1000$. The basic physical mechanism is that after air flows tangentially into the swirl chamber, it forms a spirally rotating jet flow. A balance sets up between the centrifugal forces on air molecules and the pressure forces on the chamber wall, so a low pressure region occurs in the centre of the jet flow. The low pressure in this region recovers gradually as air ascends and effuses out of the burner nozzle, thus resulting in an axial adverse pressure gradient. At high degree of swirl ($S>0.6$), the flow in the centre of the jet reverses its direction and vortices are induced. This reverse flow zone is the IRZ and its size generally increases with the degree of swirl. The presence of the IRZ leads outer boundaries of the jet to expand radially soon after its emergence from the burner nozzle, thus increasing the rate of axial velocity decay which is an important parameter influencing flame length. At the position of initial expansion, fuel is introduced into the swirling air jet flow. The initial expansion does not entrain air from the surroundings because of the confines of the divergent outlet and the supplied fuel and air mix with the confines of inner boundaries of the IRZ. Particles of fuel and air emerging from the burner nozzle may be considered to be released from the constraining force exerted by the burner wall. They would have a tendency to fly out tangentially in a manner similar to that of an object rotated at the end of a string and then released. The radial spreading of the swirling jet flow is supposed to result in a faster increase in cross-sectional area, if compared with the jet flow without induced swirl. The increased cross section is then expected to broaden the effective contact area between the flame jet and the target surface in the flame impingement heat transfer application which will be studied in Chapter 6.

Note that the degree of swirl not only depends on the relative magnitude of the supplied axial and tangential air flow rates, but also depends on the swirl generation efficiency of the swirl burner. Only part of the pressure drop through the swirl burner reappears as kinetic energy of the swirling air jet flow. The remainder is due to mechanical energy loss inside the swirl burner. This explains the reason for the existence of a threshold Reynolds number of about 1000 for the IRZ to establish. At $Re<1000$, a stable, quiet, buoyancy-driven diffusion flame is formed, without the formation of the IRZ. As Re increases beyond 1000, the angle of spread of the flame increases and the flame is observed to widen in its cross-sectional area. Meanwhile, the internal recirculation zone grows in size with the vortex centres moving away from the flame axis and shifting downstream, and the entrainment of secondary air from the surroundings increases. Note that the flame is in the laminar state at $Re < 2000$ and in the turbulent state at $Re > 2000$. The stability of the flame is maintained when Re is 20000 at which the flowmeter (Type 32121) reaches its full capacity, thus $Re=20000$ is the highest value that can be achieved. To find the highest value or upper limit of Re for a stable flame, two steps are taken. Step 1: adjust the pressure gauge for the flowmeter (Type 32121) from 4 to 7 kg/cm^2 and adjust the flowmeter to the utmost reading of 150. Step 2: Adjust the pressure gauge to 7.6 kg/cm^2 and take off the flowmeter from the tubing of air supply in order to eliminate the pressure loss caused by the flowmeter, thus we gain the highest air volumetric flow rate that can be achieved in the laboratory. However, both steps fail to find the value of Re for flame extinction and the flame is still there with combustion carrying on, making much noise. This seems to agree with the hypothesis that strongly swirling flames with recirculation can have infinite stability (Gupta *et al.*1984) .

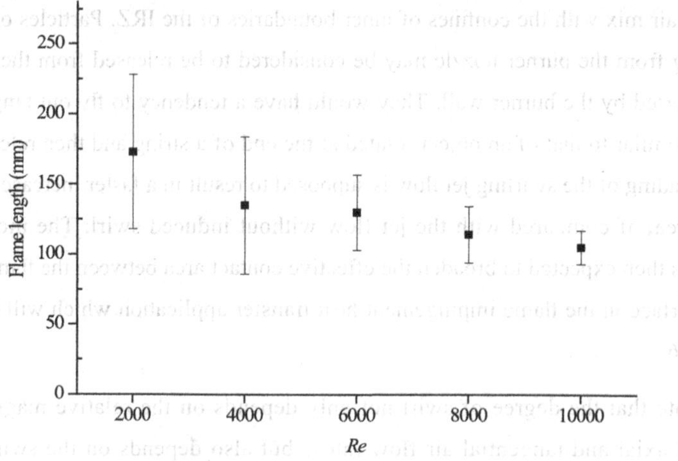

Figure 4.3 Flame images at different *Re* and *Φ*=1.5.

A quantitative analysis of the flame length was conducted and shown in Figures 4.3 and 4.4. The flame length was measured by using a digital CCD camera operating at 60 frames per second to video the luminous flame in a dark background and the flame lengths are determined by averaging a dozen images of the visible flame. Both the mean and root-mean-square visible flame lengths are shown in the figures and the RMS is shown as a bar around the mean. It is seen that the length of the swirling IDF decreases from 173 mm to 105 mm with an increase in Re from 2000 to 10000. The flame becomes mainly premixed in nature when Re is equal to or larger than 6000. This is because higher air/fuel jet velocity and turbulence level which improve air/fuel mixing and increase the decay of axial velocity component. Because of more intensive mixing, more complete combustion occurs in the IRZ of the swirling IDF.

4.1.2 Varying *Φ* at fixed *Re*

Under this circumstance the experimental tests were performed by varying the overall equivalence ratio at each Re of 1000, 2000… till to 20000. A typical case at Re=8000 with Φ increasing from 1.0 to 2.0 is shown in Figure 4.4. The increasing Φ means an increase in the amount of supplied fuel, thus the flame becomes more diffusional in nature and the flame length increases accordingly.

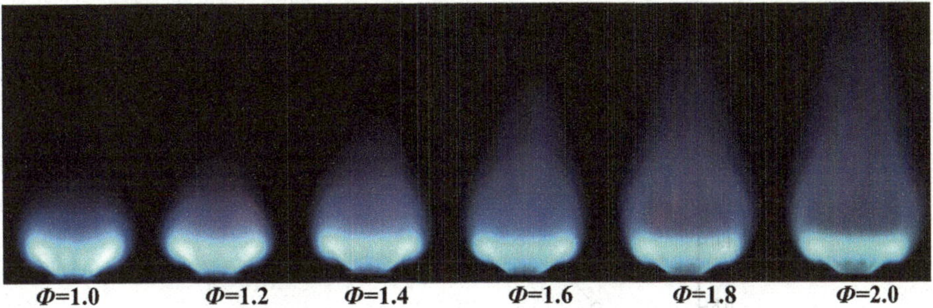

Figure 4.4 Flame images at different Φ and Re=8000.

It is seen that the swirling IDF remains fairly premixed in nature, and the variation in the shape and size of Zone 1 and Zone 2 is relatively small, compared to that of Zone 3, which augments in size when the flame length increases.

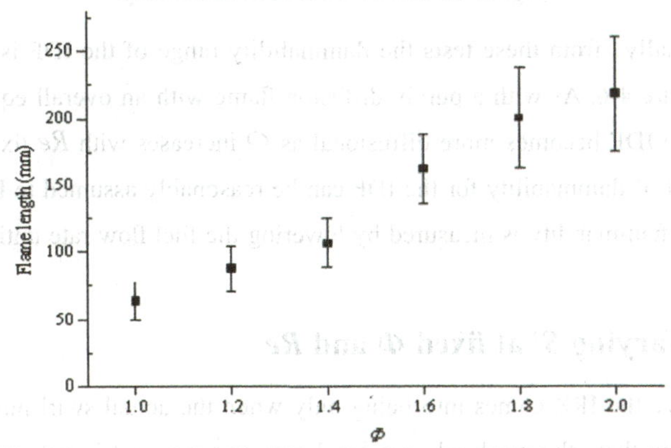

Figure 4.5 Flame length at different Φ and Re=8000.

The flame length is shown in Figure 4.5. As a result of the increasing amount of supplied fuel and due to the fact that Zone 3 or the post-combustion zone becomes more and more diffusional and thus observable as Φ increases from 1.0 to 2.0, the flame length increases monotonically.

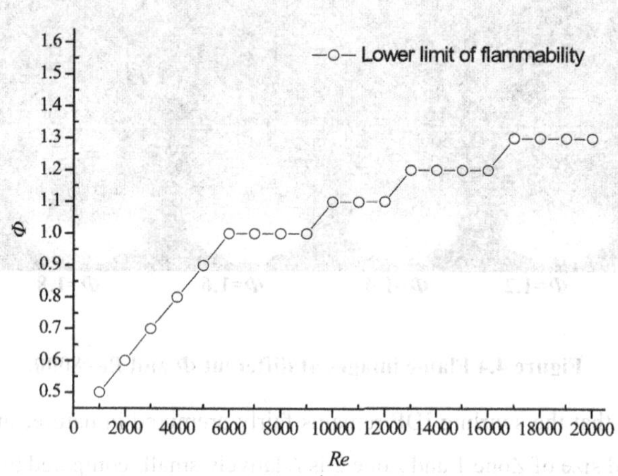

Figure 4.6 Lower limit of flammability.

Additionally, from these tests the flammability range of the IDF is identified and plotted in Figure 4.6. As with a purely diffusion flame with an overall equivalence ratio of infinity, the IDF becomes more diffusional as Φ increases with Re fixed. Therefore, the upper limit of flammability for the IDF can be reasonably assumed to be infinity. The lower limit of flammability is measured by lowering the fuel flow rate until the flame extinguishes.

4.1.3 Varying S' at fixed Φ and Re

In theory, the IRZ comes into being only when the actual swirl number is higher than 0.6. Below that, the weak adverse axial pressure gradient is not strong enough to induce a reverse flow. We found that, for our swirl burner, the threshold value of the geometric swirl number (S') for the formation of the IRZ is around 2.4. For all the swirling flames under investigation at $S'>2.4$, an IRZ is always observable if $Re>1000$. This is because the IRZ is induced by the axial adverse pressure gradient, and the axial adverse pressure gradient is dominated by the swirl number and influenced to a lesser extent by Re.

4.2 Flame structure

The identification of flame structure can be based on a couple of techniques, such as visually observing the flame, flow visualization, flame temperature distribution and in-flame species concentration. All these methods are utilized in the present study and the structure of the swirling IDF is initially characterized based on the flow field which is identified by flow visualization. Later on, the initially identified flame structure is found to be consistent with the analysis of the temperature and in-flame species concentration distributions. Hence the initially identified flame structure is confirmed to be true.

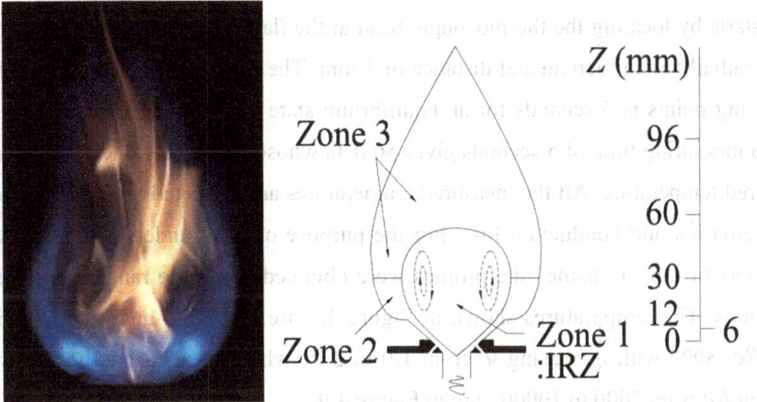

Figure 4.7 Flame photo and sketch of swirling IDF at *Re*=6000 and *Φ*=2.0.

Figure 4.7 shows the flame structure of the swirling IDF and a flame photo to the left obtained at *Re*=6000 and *Φ*=2.0. The flame is peach-shaped and consists of three zones. Zone 1 is the IRZ which is formed by the reverse flow. Zone 1 is yellow in color due to the very rich combustion occurring in this zone at *Φ*=2.0. The yellow color indicating soot formation makes the IRZ recognizable to the naked eyes. The IRZ becomes invisible when its color is sky-blue in those flames that are not so rich. In that case, the existence of the IRZ can be checked by the flow visualization technique. Zone 2 is the flame boundary in the lower section, being in contact with ambient air on the outer side and with Zone 1 on the inner side. Zone 2 is where both intense mixing of the supplied air/fuel and intense combustion occur and is always navy-blue in color due to premixed combustion. Zone 3 is the flame boundary in the upper section, most part of which has the same color as Zone 1. Zone 3 in this case is the post-combustion zone in which the remaining fuel is burned in a diffusion mode. Under the experimental conditions tested, this three-zone flame structure can be typical of all swirling flames with the formation of the IRZ. Flames operating at experimental conditions other than Re=6000 and Φ=2.0 is found to have the same flame structure.

4.3 Temperature field

The swirling IDF is axisymmetrical, thus in a half plane of the flame and at five axial distances of Z=6, 12, 30, 60 and 96 mm above the burner rim, a total of five radial temperature profiles are obtained for each operating condition of the flame. Each measurement starts by locating the thermocouple bead at the flame axis and then it is moved outwards radially at an incremental distance of 3 mm. The interval between two neighboring measuring points is 5 seconds for an equilibrium state to set up, and at each measuring point a measuring time of 5 seconds gives 50 data whose averaged value is reported as the measured temperature. All the measured temperatures are corrected to eliminate the errors from radiation and conduction loss. For the purpose of fully understanding the thermal characteristics of the flame, the profiles were obtained in a large range of experimental conditions. The temperatures shown in Figure 4.8 are measured under the condition of fixed Re=8000 with increasing Φ from 1.0 to 2.0, while those at fixed Φ=1.5 with increasing Re from 2000 to 10000, are in Figure 4.9.

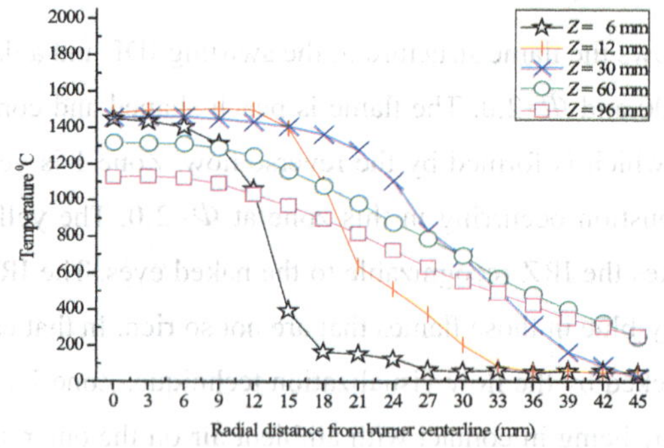

(a) Φ=1.0

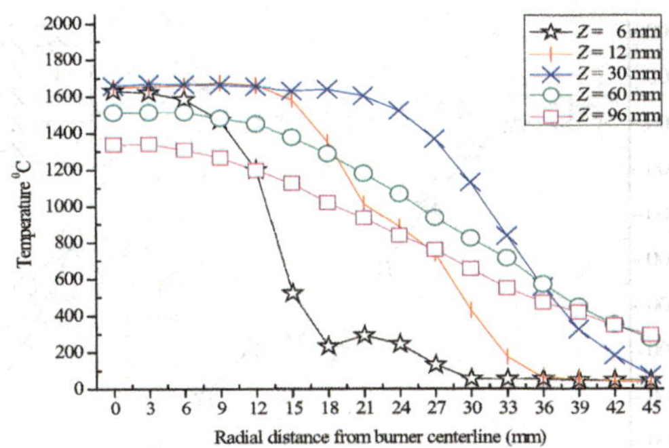

(b) *Φ*=1.2

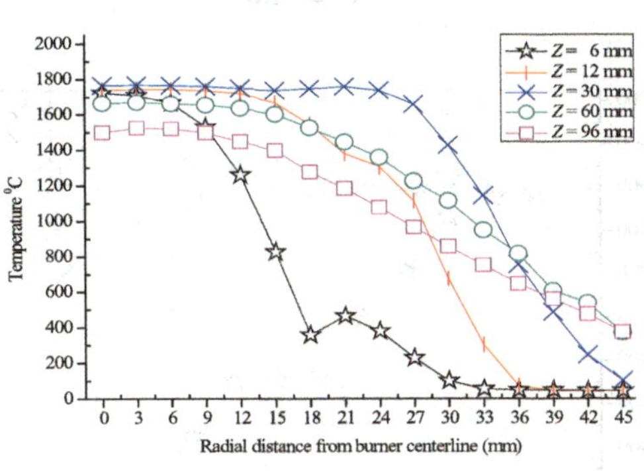

(c) *Φ*=1.4

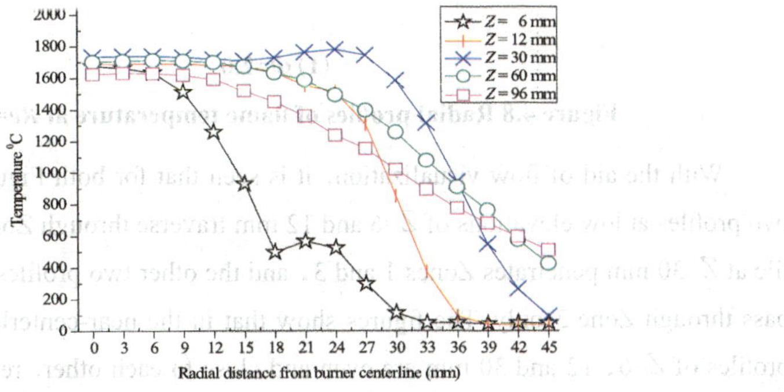

(d) *Φ*=1.6

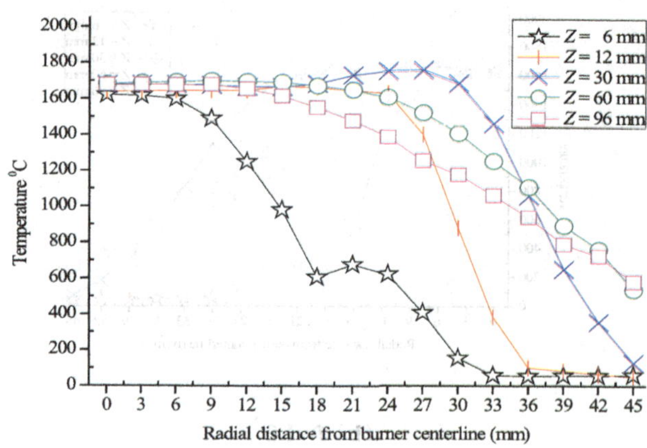

(e) Φ=1.8

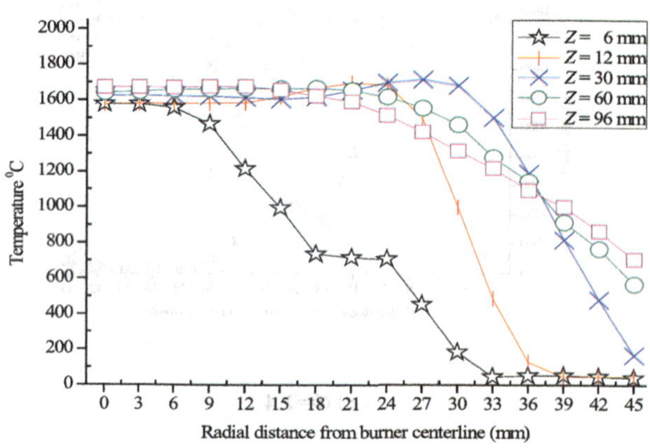

(f) Φ=2.0

Figure 4.8 Radial profiles of flame temperature at *Re*=8000.

With the aid of flow visualization, it is seen that for both Figures 4.8 and 4.9, the two profiles at low elevations of Z=6 and 12 mm traverse through Zones 1 and 2. The profile at Z=30 mm penetrates Zones 1 and 3, and the other two profiles at Z=60 and 96 mm pass through Zone 3 only. The figures show that in the near-centerline region, the three profiles of Z=6, 12 and 30 mm are even and close to each other, revealing that the temperature in Zone 1 is quite high and uniform. This is because Zone 1 is a well-stirred region where intense exothermic chemical reactions take place. For the other two profiles of

Z=60 and 96 mm, the temperature is also very high at the centerline, indicating that combustion is intense in the near-centerline region. Each profile drops outwards radially in the flame boundary and the gradient becomes steeper at lower elevations, especially in Zone 2 as specified by the two profiles of Z=6 and 12 mm. The decrease in temperature is the result of the entrainment of ambient cold air. The flame has a larger amount of entrained ambient air at higher elevations, which leads the flame boundary to thicken with elevation and thus the temperature gradient in the flame boundary becomes smaller with elevation. The profile of Z=30 mm at Φ=1.6, 1.8 and 2.0 in Figure 4.8 and the profiles of Z=12 and 30 mm in Figure 4.9 illustrate that the highest temperature is located at a certain radial distance from the burner centerline. It actually reveals that the highest temperature is formed in an annular region which is at the lower edge of the IRZ, namely the interface between Zone 1 and Zone 2. Zone 2, being an annular region enclosing the IRZ, is featured by a bright and navy-blue color and both the high temperature and the bright navy-blue color are evidence that a curved annular flame front lies at the interface between Zone 1 and Zone 2. At the edge of the IRZ, the flow velocity and the flame speed are matched and between the forward going reactants and the reverse flow of combustion products is the flame front which is associated with most intensive mixing and combustion. This finding is consistent with the statement of Schmittel *et al.* (2000) that the highest temperature is obtained at the boundary of the IRZ, and in the IRZ where mainly combustion products prevail, a low heat loss leads to higher temperatures.

Based on the fact that the profiles of Z=6, 12 and 30 mm in the near-centerline region are close to each other, it is reasonable to regard them as a whole to facilitate the analysis of the thermal characteristics of the flame. As Φ increases from 1.0 to 2.0 at fixed Re=8000, the temperature in Zone 1 first increases from around 1450℃ at Φ=1.0 to the maximum 1750℃ at Φ=1.4 and then decreases to around 1650℃ at Φ=2.0. The profiles of Z=6, 12 and 30 mm in the near-centerline region are even, so that the effect of dilution by the entrainment of ambient cold air on the temperature in Zone 1 is small and so that the temperature can be reasonably considered to be affected only by the combustion condition in Zone 1. Consequently, the variation of the temperature with increasing Φ from 1.0 to 2.0 is due to the change of the combustion condition from fuel-lean to fuel-rich in Zone 1. The highest temperature of 1750℃ at Φ=1.4 reveals that the stoichiometric combustion condition is occurring at Φ=1.4. Within the near-centerline region and at low values of Φ, the temperature is higher in Zone 1 and is progressively lower in Zone 3, indicating that exothermic combustion is basically completed in Zone 1. In contrast, at high values of Φ, the temperature is lower in Zone 1 and is higher in Zone 3, because the fuel-rich combustion condition occurs in Zone 1 and relatively more complete combustion takes place in the near-centerline region of Zone 3.

Another feature of Figure 4.8 is that the profile at each elevation has a steeper temperature gradient at higher Φ. In other words, the effect of increasing Φ is that the high-temperature zone in the centre of the flame is broadened so that the flame boundary evolves a faster change of temperature. By means of flow visualization, it is found that the size and shape of the IRZ is actually not changed in the range of $1.0<\Phi<2.0$. The reason is that the configuration of the IRZ is controlled by the swirling air jet flow or the value of *Re* and the influence of Φ is confined to the combustion condition and thus the temperature distribution.

Note that there is a small bounce in the profile of Z=6 mm in all the cases except that at Φ=1.0 in Figure 4.8. The small bounce is induced to form when the thermocouple traverses through the dent in the flame boundary or in Zone 2. The fact that the burner nozzle protrudes on the burner head causes a bluff body effect, generating an annular low-pressure region in the corner between the burner nozzle and the burner head, which finally brings a dent to the boundary of the flame. When the thermocouple moves horizontally at Z=6 mm, it goes into the dent which is ambient cold air and then back into the flame, therefore the temperature profile exhibits a small bounce.

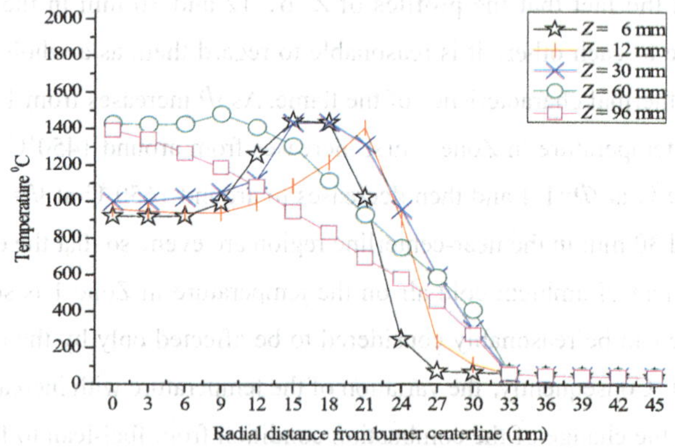

(a) *Re*=2000

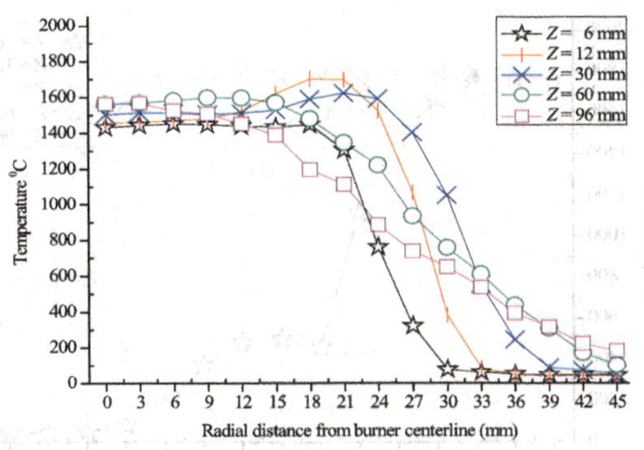

(b) *Re*=4000

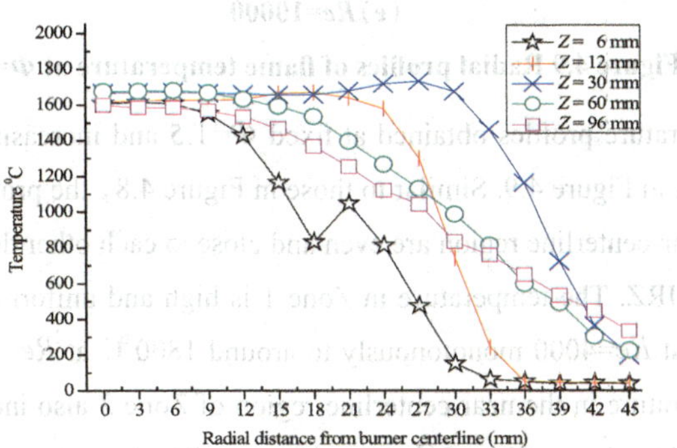

(c) *Re*=6000

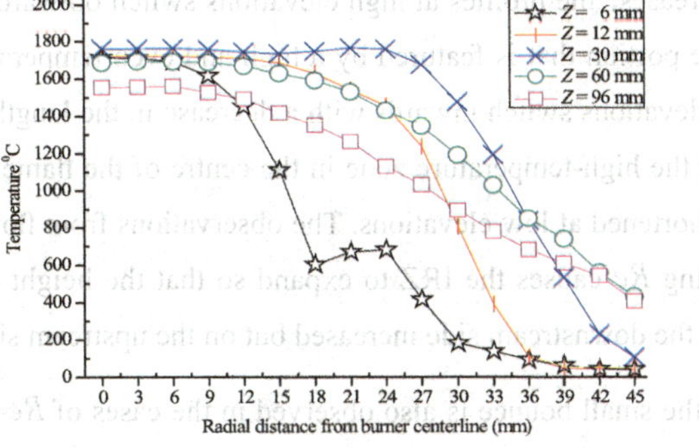

(d) *Re*=8000

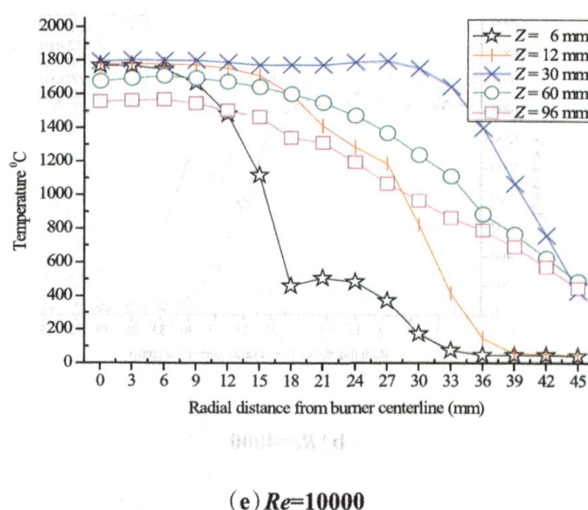

(e) ***Re*=10000**

Figure 4.9 Radial profiles of flame temperature at *Φ*=1.5.

The temperature profiles obtained at fixed *Φ*=1.5 and increasing *Re* from 2000 to 10000 are shown in Figure 4.9. Similar to those in Figure 4.8, the profiles of *Z*=6, 12 and 30 mm in the near-centerline region are even and close to each other due to the well-stirred condition in the IRZ. The temperature in Zone 1 is high and uniform and increases from around 1450℃ at *Re*=4000 monotonously to around 1800℃ at *Re*=10000. On the other hand, the temperature in the near-centerline region of Zone 3 also increases slightly with *Re*, as illustrated by the profiles of *Z*=60 and 96 mm. The increasing temperature with *Re* is simply because the higher jet flow velocity and the higher level of turbulence enhance the mixing between the supplied fuel and air and thus induce more complete combustion.

As *Re* increases, the profiles at high elevations switch outwards with an increase in the length of the portion that is featured by a high and even temperature. In contrast, the profiles at low elevations switch inwards with a decrease in the length of the even portion. This means that the high-temperature zone in the centre of the flame is broadened at high elevations but shortened at low elevations. The observations from flow visualization show that the increasing *Re* causes the IRZ to expand so that the height of the IRZ increases with it width on the downstream side increased but on the upstream side decreased.

Note that the small bounce is also observed in the cases of *Re*=6000 and *Re*=8000. For all the cases, the highest temperature along the profile of *Z*=30 mm resides at a certain radial distance from the burner centerline, confirming that the flame front is at the interface between the IRZ and Zone 2.

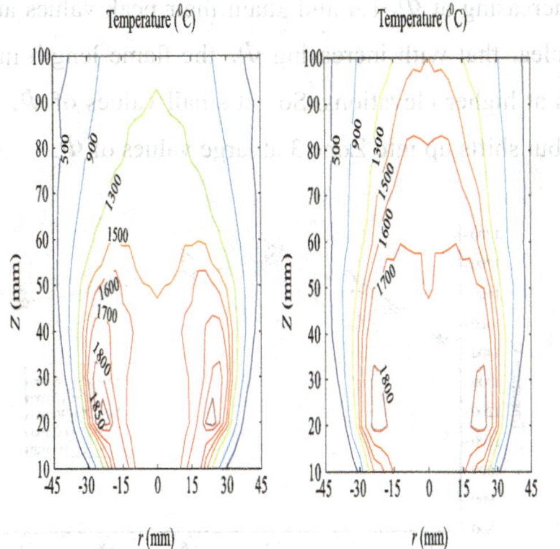

Figure 4.10 Temperature contours of the flame at *Re*=8000 and *Φ*=1.5.

The temperature field of the swirling flame operating at *Re*=8000 and *Φ*=1.5 is clearly shown in Figure 4.10. The temperature contours show that there is a high temperature in the center of the flame, with a variation of less than 200℃ in a large region which extends to around Z=50 mm and is verified by flow visualization to be the IRZ. The contour with the highest temperature is at a certain radial distance from the burner centerline, indicating that the highest temperature is produced in an annular region, consistent with the locus of the bright navy-blue flame front. The uniform, high temperature in the IRZ is due to the well-stirred reversing flow where intense exothermic chemical reactions take place. The highest temperature at the flame front is evidence that most intensive exothermic chemical reactions lies thereunto, i.e. at the edge of the IRZ.

4.4 The effect of *Re* and *Φ* on temperature

Figure 4.11 illustrates the effect of *Φ* on the centerline temperature of the swirling IDF. At *Re*=8000, the three profiles at *Z*=6, 12 and 30 mm, i.e. in Zone 1, increase with an increase of *Φ* from 1.0 to 1.4 and decrease with the further increase of *Φ* from 1.4 to 2.0. This indicates that a local complete combustion condition is established in Zone 1 at or close to *Φ*=1.4. At *Φ*>1.4, fuel-rich combustion occurs in Zone 1, leading to more intense chemical reactions occurring in Zone 3. Consequently, the profiles at *Z*=60 and

96 mm continue increasing at Φ>1.4 and attain their peak values at Φ=1.6 and Φ=1.8, respectively. It is clear that with increasing Φ, the flame length increases and the peak temperature occurs at higher elevations. So, at small values of Φ, the peak temperature occurs in Zone 1, but shifts up into Zone 3 at large values of Φ.

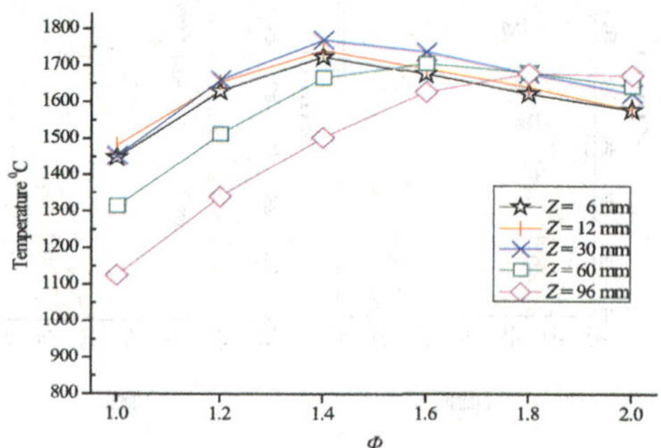

Figure 4.11 Effects of Φ on centerline temperature at *Re*=8000.

Figure 4.12 shows that with an increase of *Re* from 2000 to 10000 at Φ=1.5, the profiles in Zone 1 increase monotonically and the profiles in Zone 3 firstly increase and then tend to drop slightly. The highest temperature occurs in Zone 3 at low *Re*, but in Zone 1 at high *Re*. This is because that at higher *Re*, the size and strengthen of the IRZ are enhanced, causing a larger portion of the supplied fuel and air to mix and react in Zone 1, resulting in a higher temperature in this zone. As *Re* increases, the flame length decreases and due to the dilution of the combustion products in the post-combustion region, the temperature at *Z*=60 and 96 mm drops slightly.

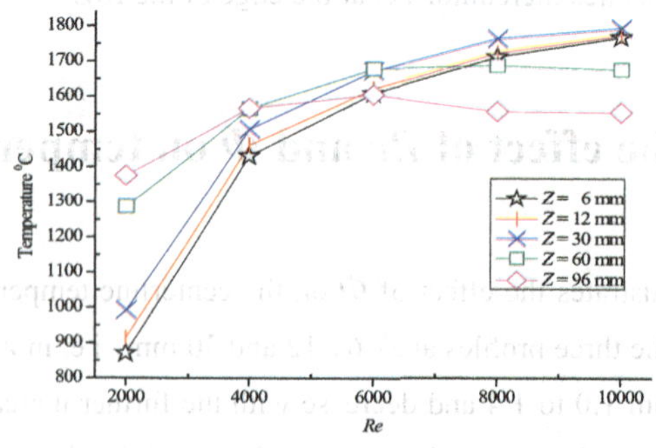

Figure 4.12 Effects of *Re* on centerline temperature at Φ=1.5.

CHAPTER 5 FLAME EMISSIONS

In Chapter 4, the swirling IDF appearance, structure and temperature have been investigated. This chapter presents the results of the experimental investigation on the emission characteristics of the flame, including both the in-flame stable gaseous species emissions and overall pollutant emissions. The concentration of the gaseous species including O_2, CO_2, CO and NO_x are measured and presented. The overall pollutants emitted by the flame are measured in the post-combustion region and presented in the form of emission indices of CO and NO_x. The effects of the Reynolds number and the overall equivalence ratio on the emission characteristics of the flame are also investigated and discussed.

5.1 In-flame stable gaseous species

The measurement of the composition of the gas within the swirling IDF was carried out and the polyatomic stable gaseous species of O_2, CO_2, CO and NO_x, which are considered to be the main constituents of the gas, are considered and their concentrations were measured by a probe sampling technique followed by in-situ analysis. For a better understanding of the relationship between the thermal and emissions characteristics of the flame, a series of simultaneous measurements of temperature and composition inside the flame are performed. Figure 5.1 shows the configuration of the measuring rig. The burner is mounted on a 3-XYZ axis stage, while the thermocouple and the sampling probe are fixed in their individual positions as shown in the figure, so when the burner translates in the X direction the thermocouple and the sampling probe alternatively touch the flame and carry out the temperature and composition measurements simultaneously without interference to each other. Each measurement is conducted by moving the burner along the X direction with an incremental distance of 3 mm. Due to the symmetry of the swirling IDF, only in a half plane of the flame and at five axial distances of Z=6, 12, 30, 60 and 96 mm above the burner rim, the temperature and concentration profiles are obtained. The data associated with the flame temperature have been presented in Chapter 4. The intervals between two successive measuring points are 60 and 20 seconds for the run of O_2 and CO measurements and for the run of NO_x and CO_2 measurements, respectively. The 60-second interval is longer than the 5-second interval for the temperature measurement because the equipment, Anapol EU-5000, requires around 60 seconds to establish an equilibrium state to display stable readings and this interval is long enough for a patch of gas sample completes its journey from the flame into Anapol and then out of Anapol, thus setting up an equilibrium state in the measuring system. The 20-second interval is chosen for the NO/NO_x analyzer and the $CO/CO_2/SO_2$ analyzer due to similar reasons. The measuring time at each measuring point is 60 seconds for both runs of the measurements, thus generating 60 data which are averaged to be the reported concentration.

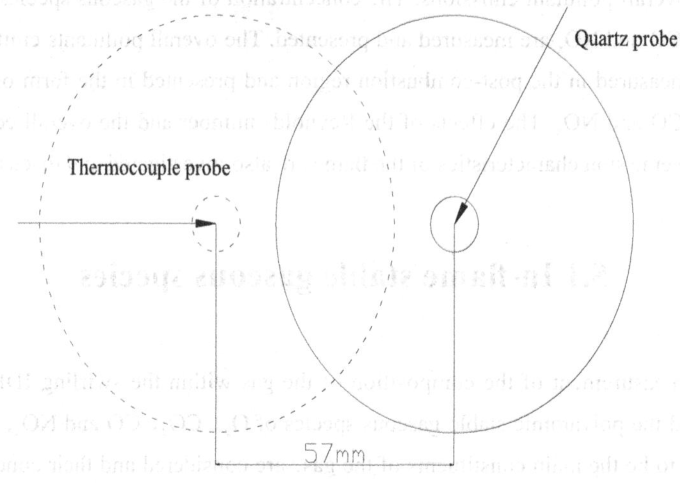

Figure 5.1 Configuration of the rig for simultaneous measurements of gas temperature and composition.

5.1.1 O_2 concentration

The radial profiles of the O_2 concentration measured at fixed *Re*=8000 with increasing *Φ* from 1.0 to 2.0 are shown in Figure 5.2. From each of these six cases shown in the figure, it can be seen that the profiles of *Z*=6, 12 and 30 mm which traverse the IRZ are steady and close to each other in the near-centreline region, indicative of a uniform distribution of the O_2 concentration, because the reversing flow in the IRZ induces a strong mixing between the air, fuel and combustion products. Each profile of the O_2 concentration is the lowest close to the centreline, which shows that intensive combustion takes place in the near-centreline region of the flame, and it increases radially outwards. Moreover, close to the centreline the O_2 concentration increases with elevation. The radial increase is due to the mixing between the entrained ambient air and the combustion products in the flame boundary. The latter increase is because a larger amount of entrained ambient air occurs at higher elevations, thus generating a better dilution of the combustion products.

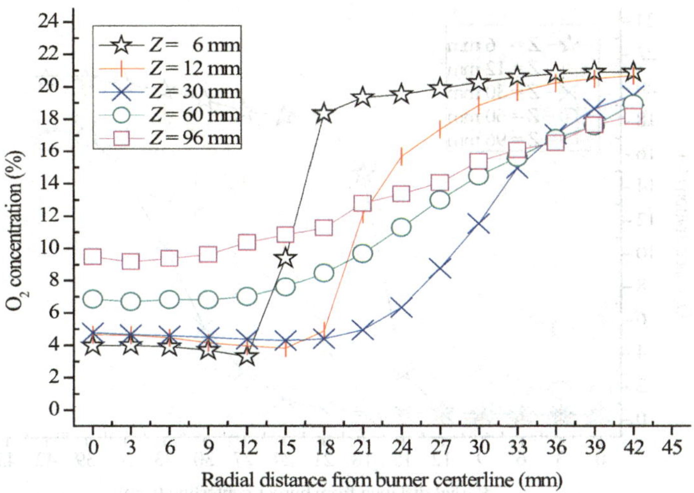

(a) Φ=1.0

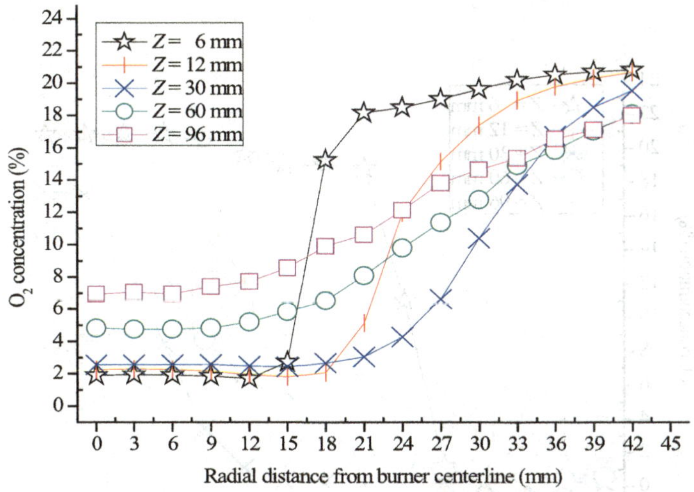

(b) Φ=1.2

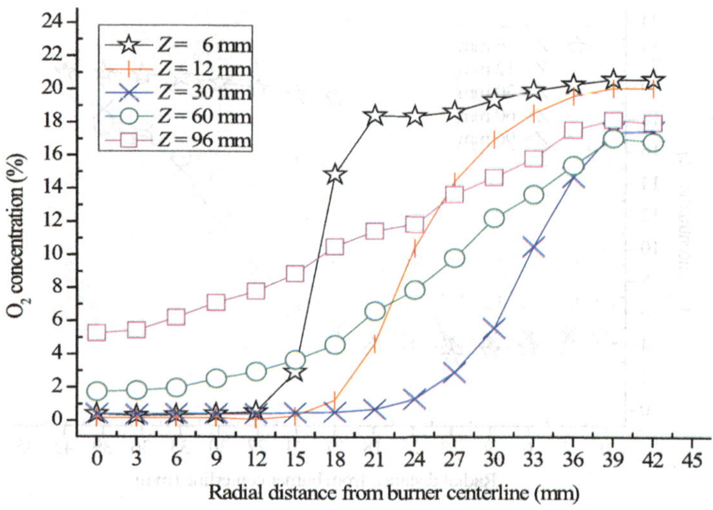

（c）Φ=1.4

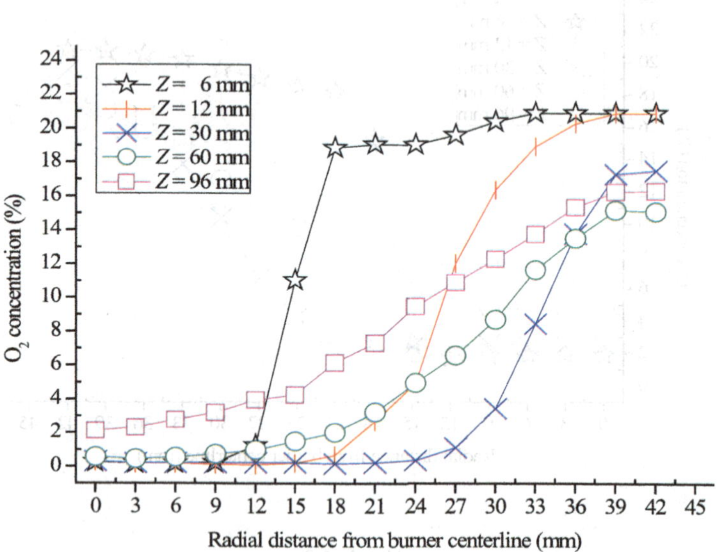

（d）Φ=1.6

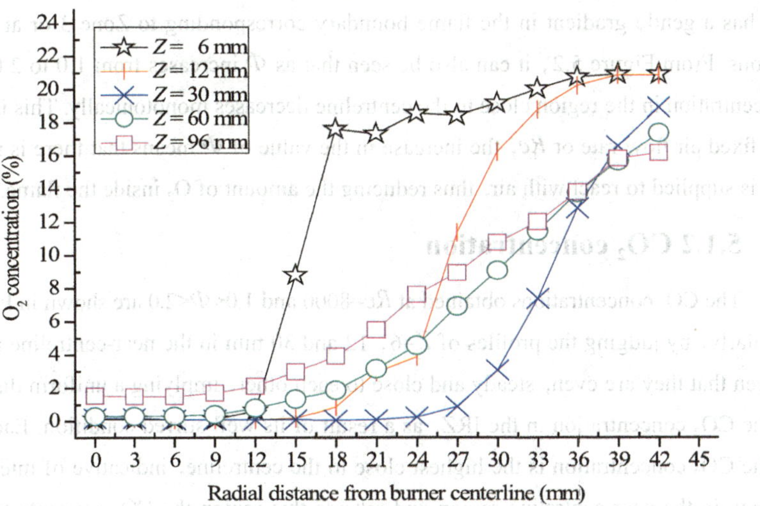

(e) Φ=1.8

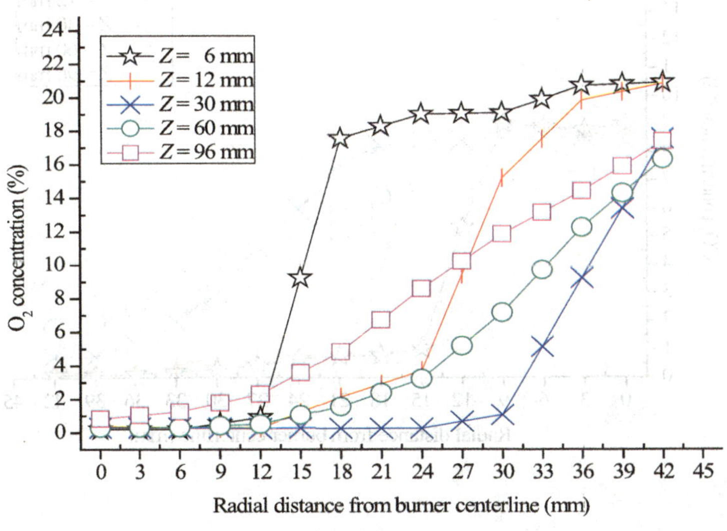

(f) Φ=2.0

Figure 5.2 Radial profiles of O_2 concentration at *Re*=8000.

Note that the O_2 concentration has a steep gradient in Zone 2 or at low elevations, and has a gentle gradient in the flame boundary corresponding to Zone 3 or at high elevations. From Figure 5.2, it can also be seen that as Φ increases from 1.0 to 2.0, the O_2 concentration in the region close to the centreline decreases monotonically. This is because at a fixed air flow rate or *Re*, the increase in the value of Φ means that there is more fuel that is supplied to react with air, thus reducing the amount of O_2 inside the flame.

5.1.2 CO_2 concentration

The CO_2 concentrations obtained at Re=8000 and $1.0<\Phi<2.0$ are shown in Figure 5.3. Similarly, by judging the profiles of Z=6, 12 and 30 mm in the near-centreline region, it is seen that they are even, steady and close to each other, implying a uniform distribution of the CO_2 concentration in the IRZ, as a result of its well-stirred condition. Each profile of the CO_2 concentration is the highest close to the centreline, indicative of intense combustion in the near-centreline region and outside this region the CO_2 concentration drops radially outwards due to the effect of dilution caused by entrained ambient air.

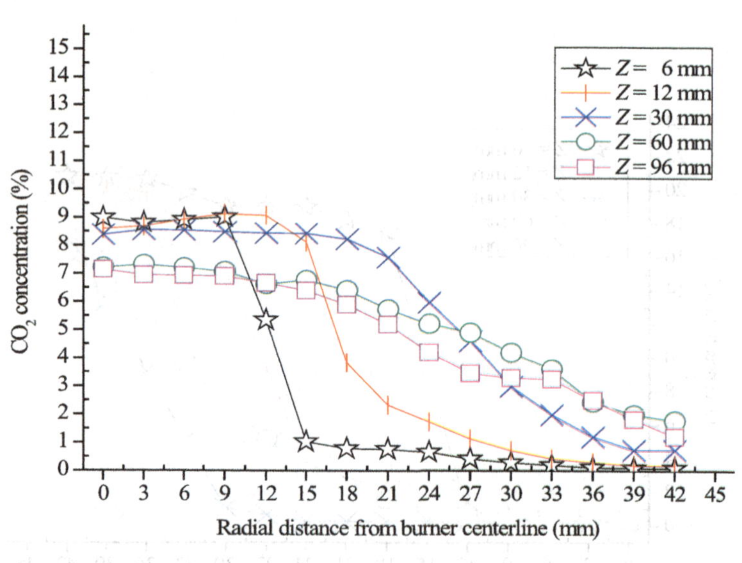

(a) Φ=1.0

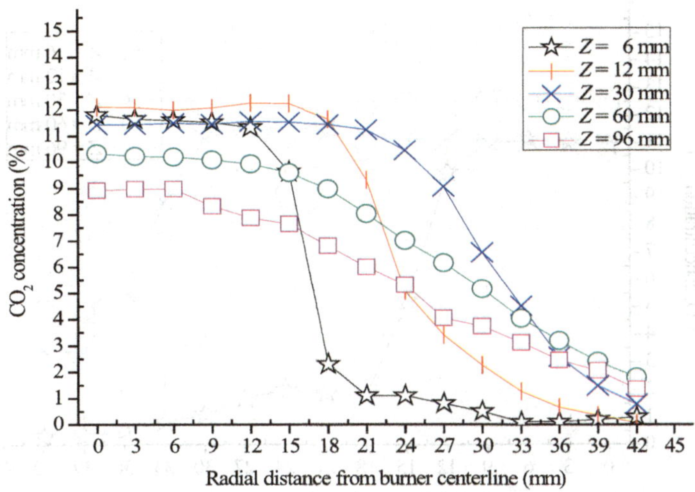

(b) Φ=1.2

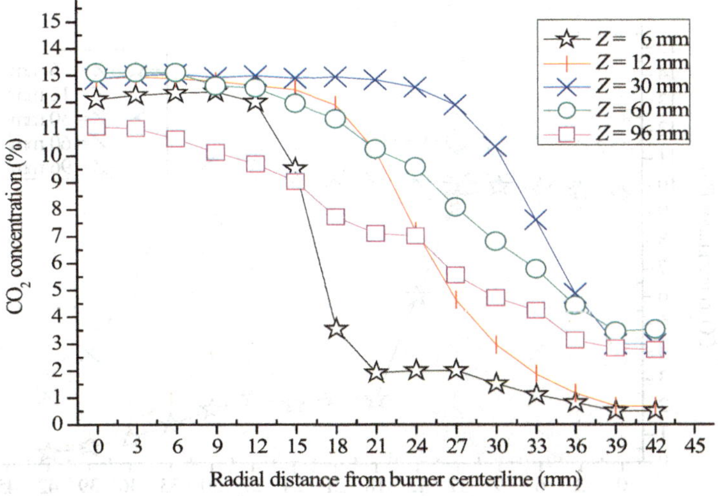

(c) Φ=1.4

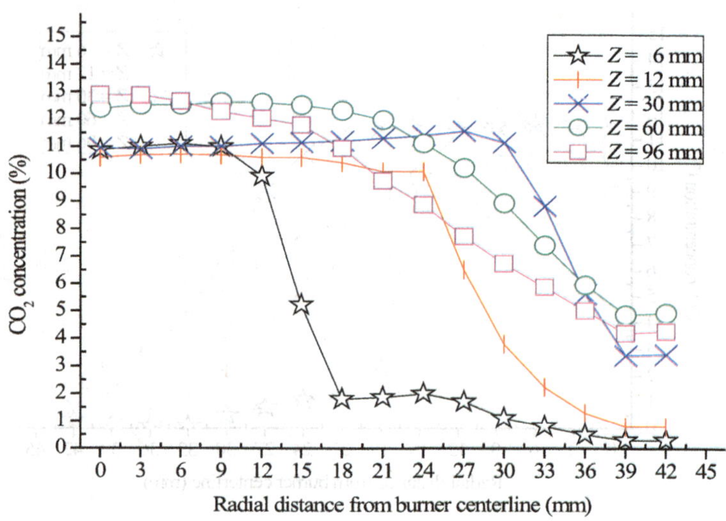

（d）Φ=1.6

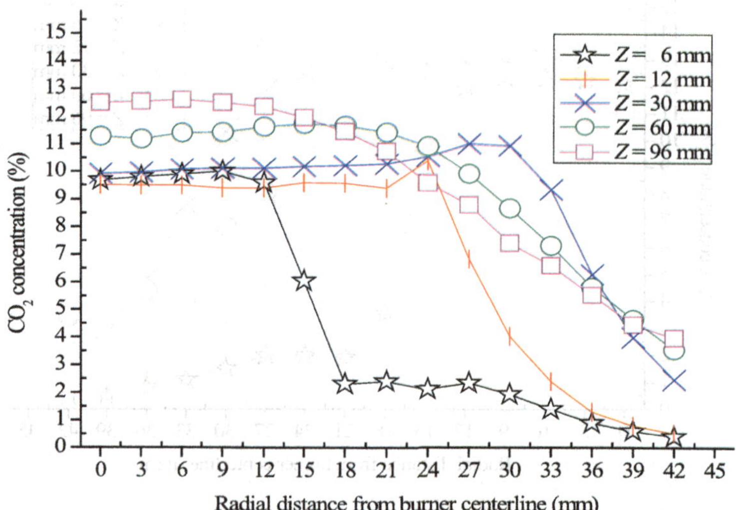

（e）Φ=1.8

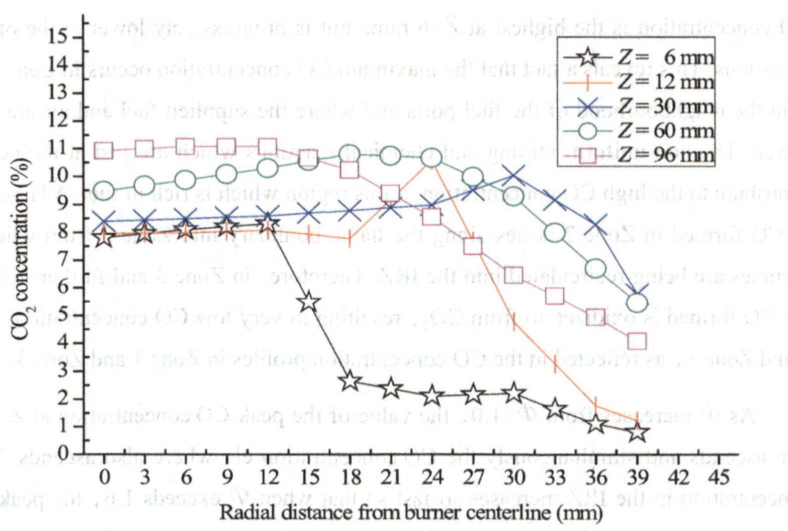

(f) Φ=2.0

Figure 5.3 Radial profiles of CO_2 concentration at *Re*=8000.

There is an upsurge of CO_2 at a certain radial position close to the flame front, where most of the chemical reactions take place. As Φ increases from 1.0 to 2.0 at Re=8000, it is seen that near to the centreline, the CO_2 concentration is higher in Zone 1 than that in Zone 3 at $\Phi \leq 1.4$, implying that combustion is more intense and goes to completion in Zone 1 and that combustion products are slightly diluted by atmospheric air in Zone 3. However, in the near-centreline region the CO_2 concentration is higher in Zone 3 than that in Zone 1 at Φ>1.4, meaning that more intense combustion occurs and more combustion products accumulate in Zone 3.

5.1.3 CO concentration

Figure 5.4 shows the radial profiles of the CO concentration at Re=8000 with increasing Φ from 1.0 to 2.0. Similarly, a uniform distribution of CO is observed again in the IRZ. All the profiles are very low in the near-centreline region, and those in Zone 1 higher than those in Zone 3. Close to the centreline, the CO concentration decreases with elevation, to the contrary of the variation of the O_2 concentration, as O_2 and CO react with each other to form CO_2. The profiles of Z=6 and 12 mm in the four cases at $\Phi \leq 1.6$ and the profiles of Z=6, 12 and 30 mm in the three cases at $\Phi \leq 1.4$ show that the CO concentration ascends to a peak value and then descends radially outwards. In each case, the peak CO concentration is the highest at Z=6 mm, but is progressively lower at the other two elevations. This reveals a fact that the maximum CO concentration occurs in Zone 2 which is in the neighbourhood of the fuel ports and where the supplied fuel and air are initially mixed. The non-uniform mixing and chemical reactions which are just at the beginning contribute to the high CO concentration in this region which is rich in fuel. A large portion of CO formed in Zone 2 issues along the flame boundary into Zone 3 from where fluid particles are being recirculated into the IRZ. Therefore, in Zone 3 and further in Zone 1, the CO formed is oxidized to from CO_2, resulting in very low CO concentrations in Zone 1 and Zone 3, as reflected in the CO concentration profiles in Zone 1 and Zone 3.

As Φ increases from Φ=1.0, the value of the peak CO concentration at Z=6 or 12 mm ascends and simultaneously the CO concentration elsewhere also ascends. The CO concentration in the IRZ increases so fast so that when Φ exceeds 1.6, the peak at Z=6 or 12 mm becomes indiscernible for that the CO concentration in the IRZ is of the same magnitude with that in the neighbouring region of the fuel ports. This is evidence that most of the fuel flowing out of the fuel ports is reversed back into the IRZ, thus leading the incomplete combustion product of CO to accumulate in the IRZ at a fast rate.

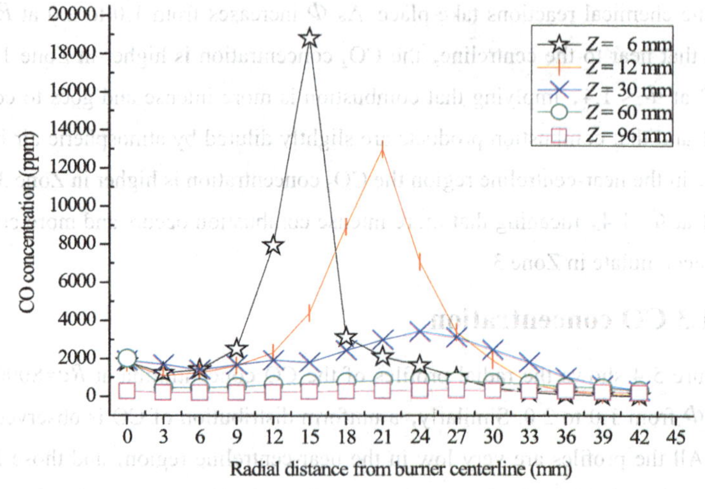

(a) Φ=1.0

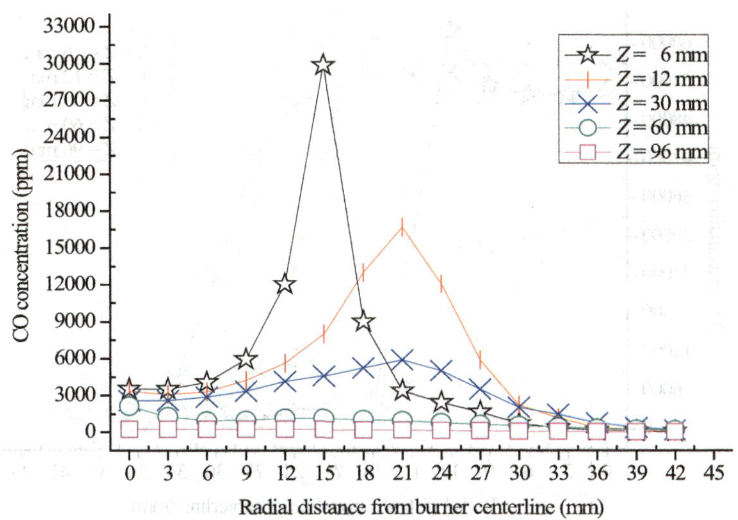

(b) Φ=1.2

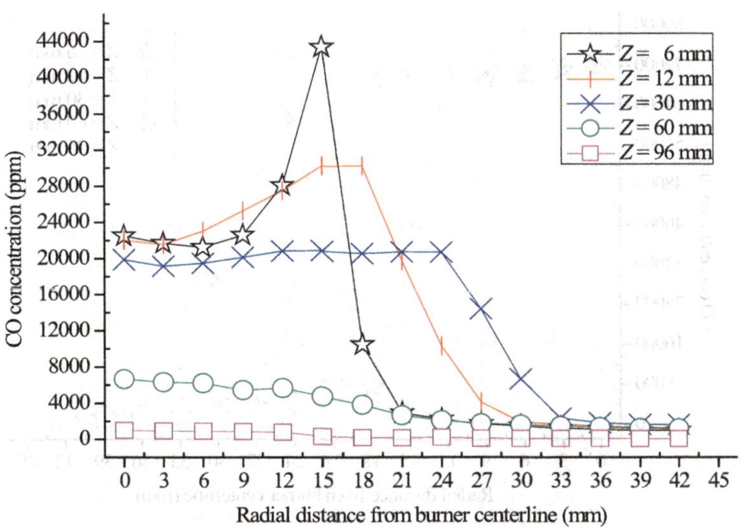

(c) Φ=1.4

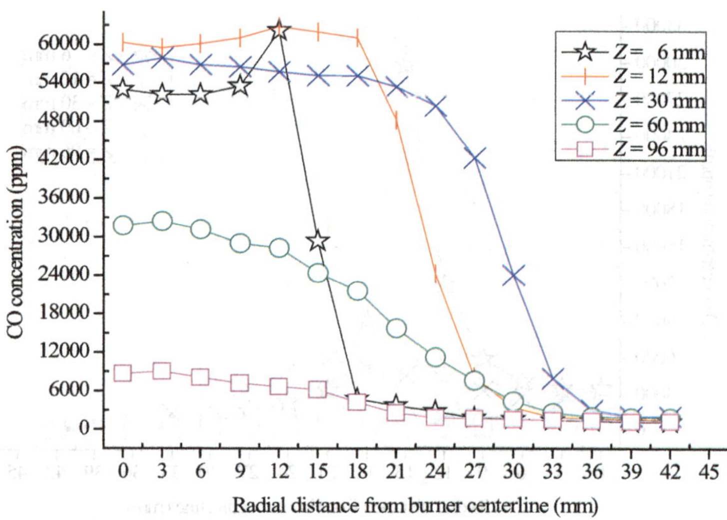

（d）Φ=1.6

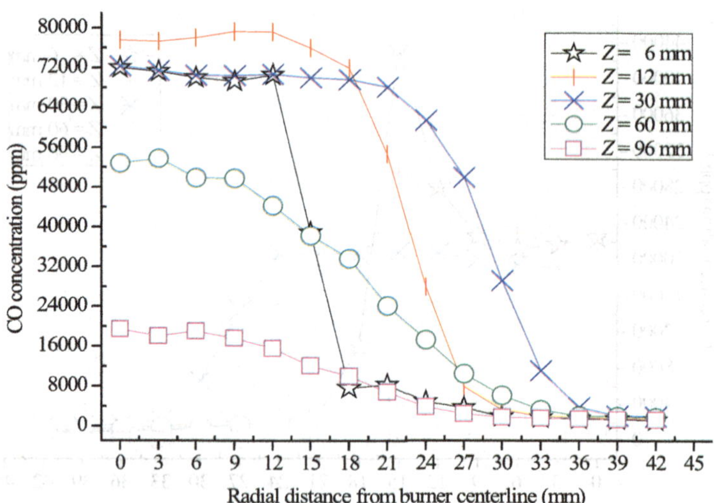

（e）Φ=1.8

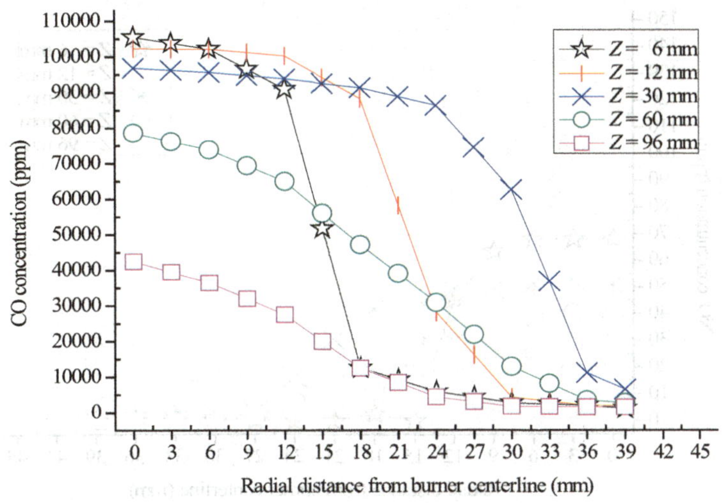

(f) Φ=2.0

Figure 5.4 Radial profiles of CO concentration at *Re*=8000.

What is more, with increasing Φ, the CO concentration elsewhere also increases due to the bigger amount of supplied fuel. Also, it is obvious that the mixing between the supplied fuel and air is mainly located in Zone 2 because of its high CO concentration for any value of Φ. Zone 1 is the second place for the mixing and at large values of Φ, more fuel is reversed back into Zone 1 for further mixing between the supplied fuel and air, based on the movement of the peak CO concentration from Zone 2 to Zone 1.

5.1.4 NO_x concentration

The NO_x concentration profiles at *Re*=8000 and $1.0<\Phi<2.0$ are shown in Figure 5.5. It is seen that the NO_x concentration distribution resembles the corresponding CO_2 concentration distribution. Moreover, the CO_2 concentration distribution is found to be similar to the temperature distribution.

To better analyze the formation mechanism of NO and NO_2 in the flame, the species concentration distributions and the temperature distribution are put together into Figure 5.6 for the case of *Re*=8000 and Φ=1.2.

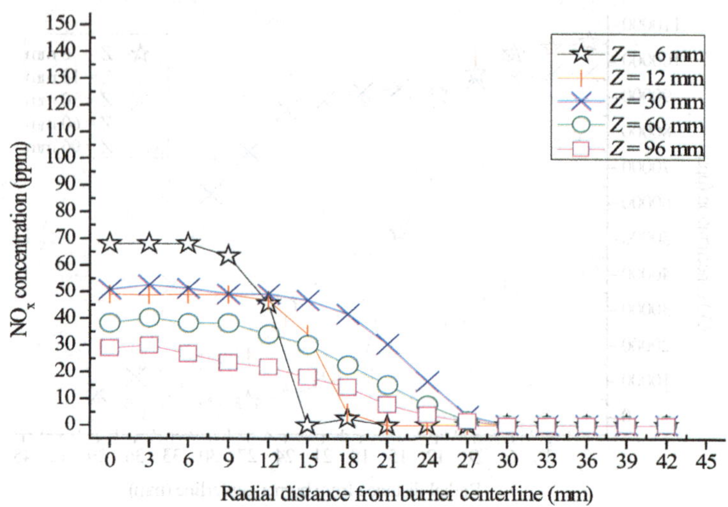

(a) Φ=1.0

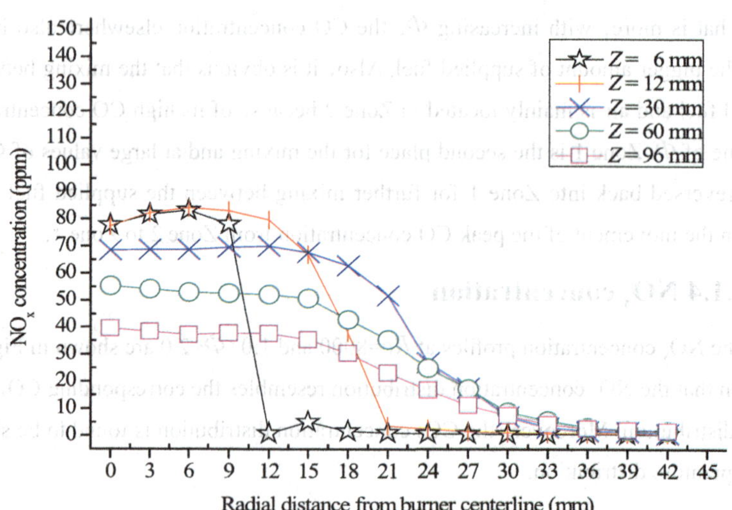

(b) Φ=1.2

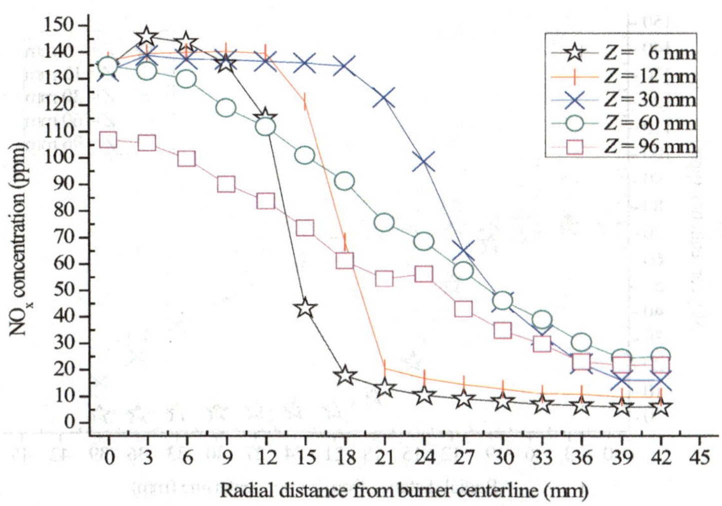

(c) Φ=1.4

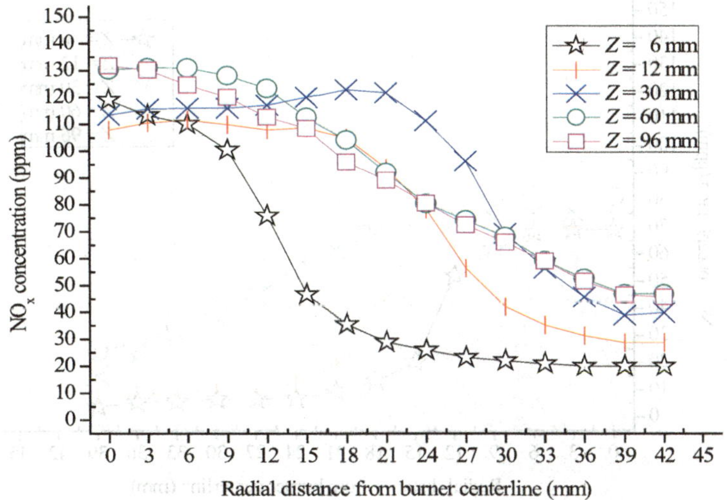

(d) Φ=1.6

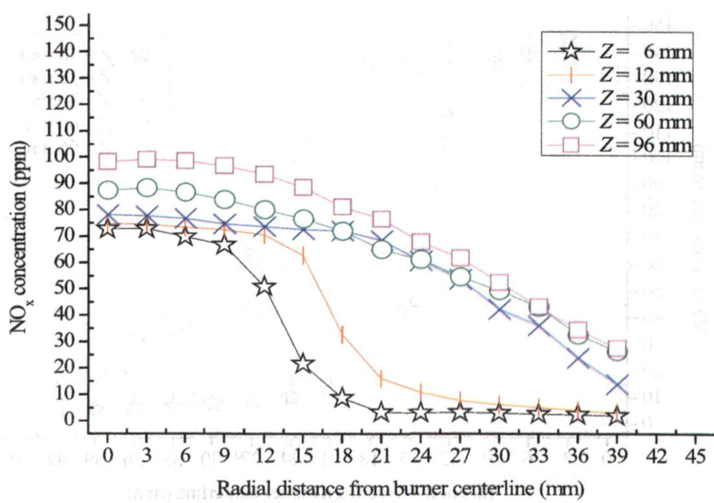

(e) Φ=1.8

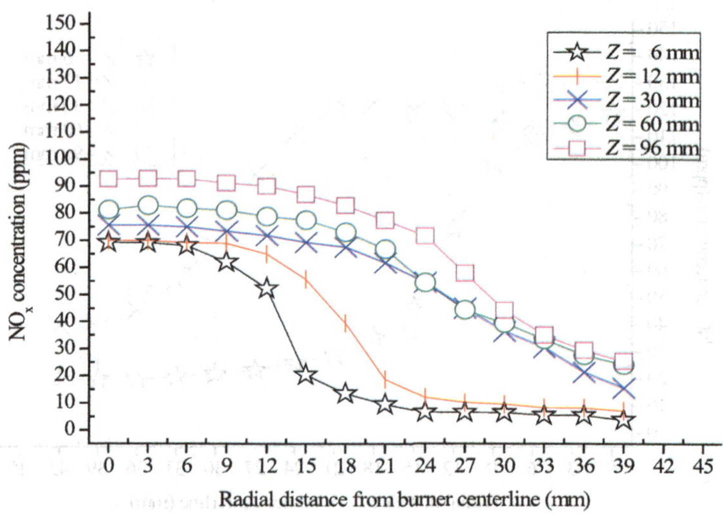

(f) Φ=2.0

Figure 5.5 Radial profiles of NO_x concentration at *Re*=8000.

The analysis of the relationship of these distributions shows that the NO_x profiles resemble the corresponding CO_2 and temperature profiles. The CO_2 and temperature profiles vary in the same directions because heat is released in the combustion process which leads to the formation of CO_2. Consequently, it derives that the formation of NO increases with temperature and the figure shows that the NO_x concentration decreases very rapidly when the temperature is below about 1800 K. This is consistent with a frequently applied rule-of-thumb that the thermal formation of NO plays a greater role in the overall NO formation for high-temperature flames above 1900 K (William *et al.*1999). The quantity of NO produced via the prompt NO mechanism is not appreciable in the present case of Re=8000 and Φ=1.2, because the NO_x concentration in Zone 2, which has fuel-rich combustion illustrated by the peaks in the CO concentration profiles, are relatively low. The formation of NO at Z=60 and at E 96 mm occurs probably due to the contribution of the residence time, which becomes longer in Zone 3.

The species concentrations have sharp gradients in Zone 2 or at low elevations and gentle gradients in Zone 3 or at high elevations. Zone 2 has fast changes not only in concentrations of the gaseous species but also in temperature. It is the fluid dynamics that dominates. Syred *et al.* (1971) reported that the streamlines in the recirculation zone are sparse and those in the boundary close to the burner rim are dense. Thus the local flow velocity, pressure, and flux change faster in the region close to the burner rim than those in the recirculation zone. The swirling air carrying with it the fuel gas forms a narrow annular boundary close to the burner rim, namely forming Zone 2 and then Zone 3. Flow separation occurs under the effect of axial adverse pressure gradient. Therefore a portion of the mixture in Zone 3 is scratched into the central region and is reversed back towards the burner nozzle. The air/fuel mixture behaves like being squeezed into the narrow annulus and then released into the relatively large IRZ. Consequently, sharp gradients in temperature and species concentrations emerge in Zone 2.

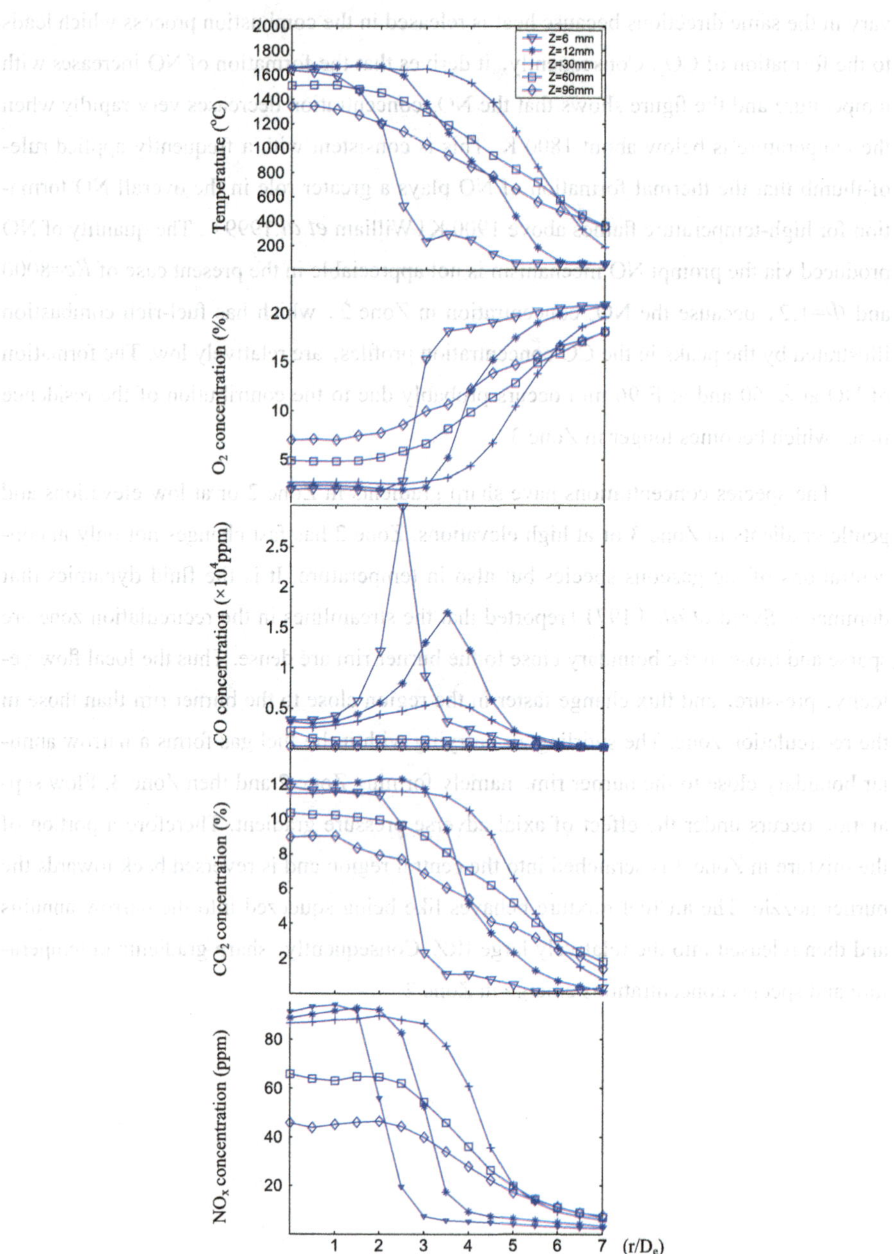

Figure 5.6 Temperature and species distributions at *Re*=8000 and *Φ*=1.2.

5.2 The effects of *Re* and *Φ*

The effects of *Re* and *Φ* on O_2, CO, CO_2 and NO_x concentrations along the centerline of the swirling IDF are shown in Figures 5.7 – 5.14. The effect of *Φ* is investigated at *Re*=8000 while the effect of *Re* is investigated at *Φ*=1.5. From the figures, it can be observed that the centerline concentration profiles for each species in Zone 1 are very close to each other, indicating that the air/fuel mixture, combustion products and entrained ambient air are well mixed in this zone.

5.2.1 O_2 concentration

From Figure 5.7, it is seen that there is a strong dependence of the centerline O_2 concentration on *Φ* at *Re*=8000. It is seen that as *Φ* increases from 1.0 to 2.0, the O_2 concentration drops at all locations due to the increasing fuel supply. At *Φ*=1.0 and 1.2, the centerline O_2 concentrations are more than 0.9% even at *Z*=6 mm, indicating that the flame is basically fuel-lean in nature at these two overall equivalence ratios. At higher *Φ*, the flame becomes progressively longer, leading to much lower O_2 concentrations even in the two measuring locations of *Z*=60 and 96 mm in Zone 3. At high level of *Φ*, the O_2 concentration is very low in Zone 1, indicating that the oxygen is nearly fully consumed, and combustion becomes fuel-rich in the IRZ.

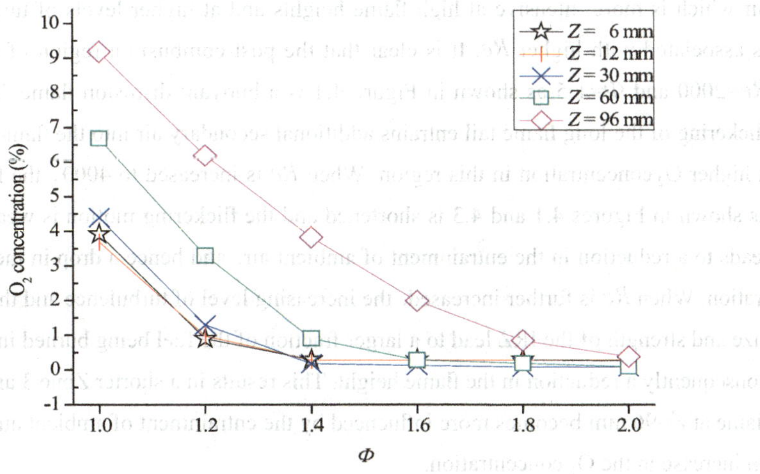

Figure 5.7 Effects of *Φ* on centerline O_2 concentration at *Re*=8000.

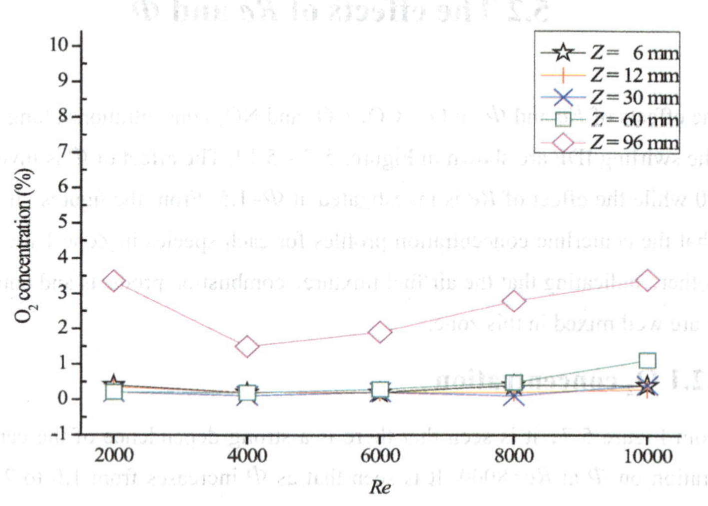

Figure 5.8 Effects of *Re* on centerline O_2 concentration at Φ=1.5.

From Figure 5.8, it can be seen that the centerline O_2 concentration in Zone 1 is almost independent of *Re* and is quite low. At *Z*=60 mm, the centerline O_2 concentration increases slightly from 0.2% to 1.1% with an increase of *Re* from 2000 to 10000. At *Z*=96 mm, the O_2 concentration is quite high and has a minimum value of about 1.5% at *Re*=4000. The higher O_2 concentration in Zone 3 is due to the entrainment of ambient air which is more intensive at high flame heights and at higher levels of turbulence that is associated with higher *Re*. It is clear that the post-combustion region of the flame at *Re*=2000 and Φ=1.5 as shown in Figure 4.1 is a buoyant diffusion flame. The buoyant flickering of the long flame tail entrains additional secondary air into the flame, leading to a higher O_2 concentration in this region. When *Re* is increased to 4000, the flame length as shown in Figures 4.1 and 4.3 is shortened and the flickering motion is weakened. This leads to a reduction in the entrainment of ambient air, and hence a drop in the O_2 concentration. When *Re* is further increased, the increasing level of turbulence and the increasing size and strength of the IRZ lead to a larger fraction of the fuel being burned in Zone 1 and consequently a reduction in the flame height. This results in a shorter Zone 3 and hence the flame at *Z*=96 mm becomes more influenced by the entrainment of ambient air, leading to an increase in the O_2 concentration.

5.2.2 CO_2 concentration

In Figure 5.9, at *Re*=8000, the centerline CO_2 concentration in Zone 1 firstly increases to a peak value at Φ=1.4, and then decreases as Φ further increases to 2.0. In Zone 3, the CO_2 concentration is lower than that in Zone 1 at low Φ but exceeds that in Zone 1 at higher Φ. In Zone 3, the CO_2 concentration attains its peak value at a slightly higher value of Φ=1.6. In Zone 1, a fuel-lean combustion condition switches to a complete combustion condition with an increase of Φ from 1.0 to 1.4, leading to an increase in the CO_2 concentration in this Zone. A fuel-rich combustion condition appears at Φ>1.4 and the CO_2 concentration begins to drop due to incomplete combustion. Combustion is completed at higher flame heights, leading to higher CO_2 concentrations in Zone 3, indicated by the peak CO_2 concentrations at Φ=1.6 for the profiles of Z=60 and 96 mm. With an increase of Φ from 1.6 to 2.0, the CO_2 concentration at Z=60 and 96 mm decreases, indicating that combustion is also incomplete at these flame heights and the location of the peak CO_2 concentration has shifted upwards or downstream to higher elevations.

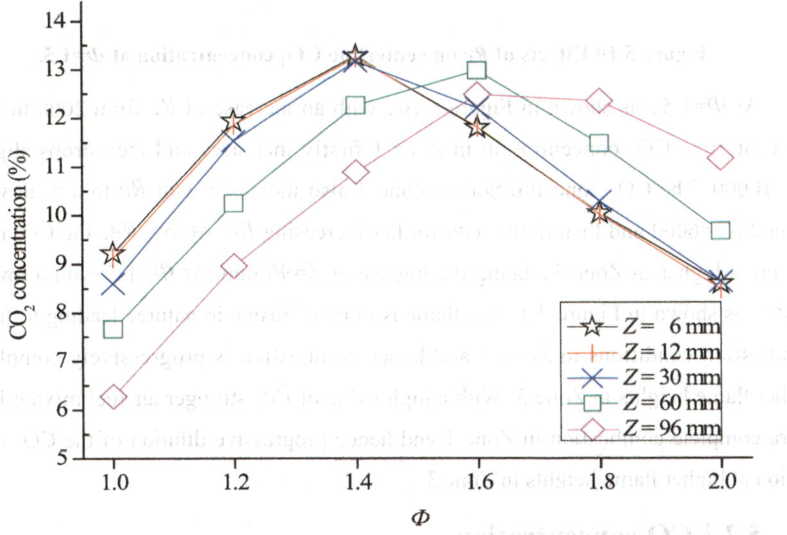

Figure 5.9 Effects of Φ on centerline CO_2 concentration at *Re*=8000.

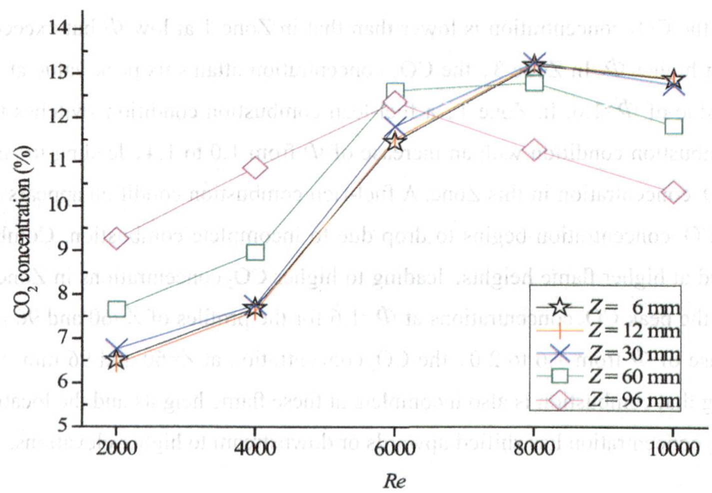

Figure 5.10 Effects of *Re* on centerline CO_2 concentration at Φ=1.5.

At Φ=1.5, as shown in Figure 5.10, with an increase of *Re* from 2000 to 10000, the centerline CO_2 concentration in Zone 1 firstly increases and then drops slightly at *Re*=10000. The CO_2 concentration in Zone 3 also increases with *Re* to a peak value at around *Re*=6000 and then drops with further increasing *Re*. At low *Re*, the CO_2 concentration is higher in Zone 3, being the highest at *Z*=96 mm. At Φ=1.5 with a low value of *Re*, as shown in Figure 4.1, the flame is more diffusive in nature, leading to fuel-rich combustion conditions in Zone 1 and hence combustion is progressively completed at higher flame heights in Zone 3. With a high value of *Re*, stronger air/fuel mixing leads to more complete combustion in Zone 1 and hence progressive dilution of the CO_2 concentration at higher flame heights in Zone 3.

5.2.3 CO concentration

Figure 5.11 shows that the centerline CO concentrations are higher in Zone 1 than those in Zone 3, probably due to further oxidation of CO to CO_2 as well as dilution by entrained air. At fixed *Re* of 8000, the CO concentration increases with an increase in Φ, especially at Φ>1.2; and at fixed Φ of 1.5, as shown in Figure 5.12, it decreases with an increase of *Re* from 2000 to 10000, especially at *Re*>4000. The increasing level of the CO concentration with Φ indicates an increasing level of fuel-rich combustion as Φ is increased to 1.4 and beyond.

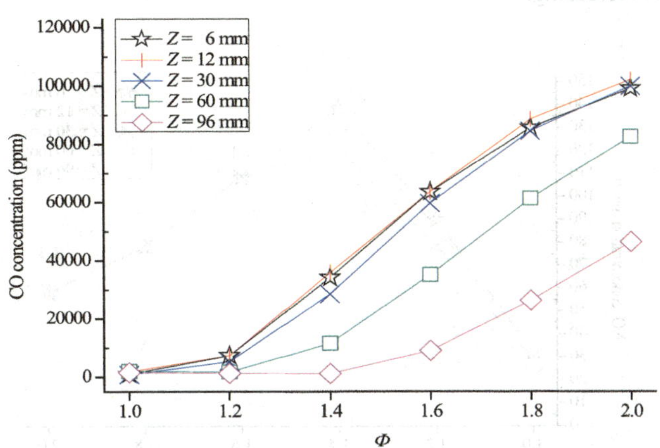

Figure 5.11 Effects of Φ on centerline CO concentration at *Re*=8000.

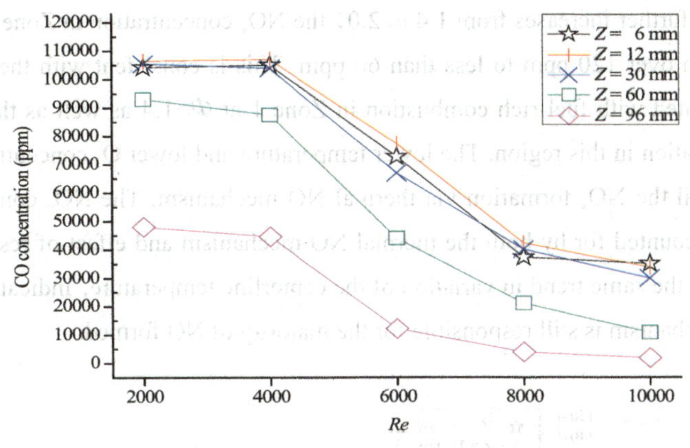

Figure 5.12 Effects of *Re* on centerline CO concentration at Φ=1.5.

In additian , as *Re* is increased, the increasing level of mixing and turbulence enhances combustion as well as the atmospheric air entrained to support combustion, leading to lower CO concentrations in the flame.

5.2.4 NO_x concentration

The centerline NO_x concentrations are shown in Figure 13. When Φ increases from 1.0 to 1.4 at *Re*=8000, there is a rise in the NO_x concentration in Zone 1, which is consistent with the increasing centerline temperature, indicating that the production rate of thermal NO is increasing.

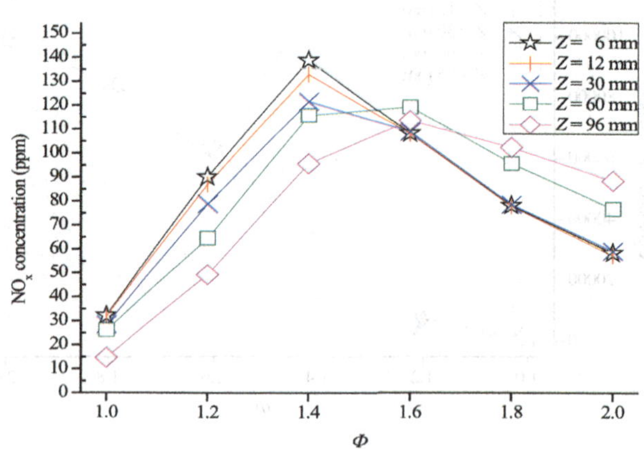

Figure 5.13 Effects of Φ on centerline NO_x concentration at *Re*=8000.

As Φ further increases from 1.4 to 2.0, the NO_x concentration in Zone 1 drops very rapidly from over 120 ppm to less than 60 ppm. This is consistent with the temperature drop associated with fuel-rich combustion in Zone 1 at Φ>1.4 as well as the decreasing O_2 concentration in this region. The lower temperature and lower O_2 concentration significantly curtail the NO_x formation via thermal NO mechanism. The NO_x concentration in Zone 3, accounted for by both the thermal NO mechanism and effect of residence time, also follows the same trend in variation of the centerline temperature, indicating that thermal NO mechanism is still responsible for the majority of NO formed.

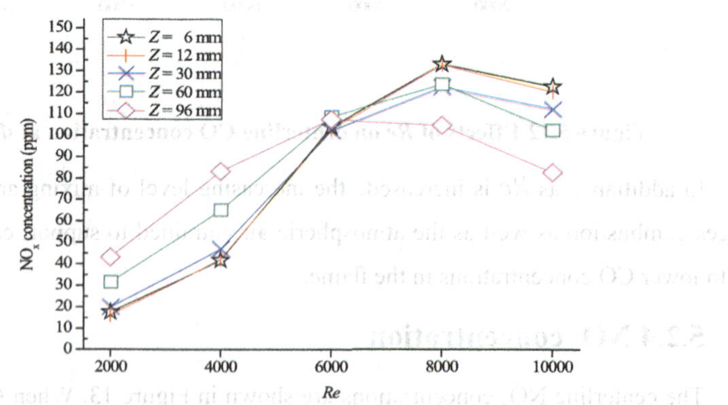

Figure 5.14 Effects of *Re* on centerline NO_x concentration at Φ=1.5.

Figure 5.14 shows that when *Re* is increased from 2000 to 8000 at Φ=1.5, the centerline NO_x concentration in Zone 1 follows the increasing centerline temperature, indicating the dominance of thermal NO mechanism. When *Re* is further increased from 8000 to 10000, a reduction in the NO_x concentration occurs, despite an increase in the temperature and O_2 concentration. The reason might be that the residence time available for NO_x formation is reduced, because of a higher flow rate associated with higher *Re* and a correspondingly smaller flame length. The same reason explains the variation in the NO_x concentration in Zone 3.

5.3 Overall pollutant emissions

The overall pollutant emissions characteristics of the swirling IDF are quantified by the emission index, expressed in grams of pollutant emitted per kilogram of fuel burned. The calculation of the emission index requires the simultaneous measurement of the pollutant and CO_2 in the flue gas. Our interest is put on the pollutants of NO_x and CO. The term NO_x typically refers to any binary compound of oxygen and nitrogen or to a mixture of such compounds, and is a major gaseous pollutant contributing the ozone depletion and acid rain. CO is a colorless, odorless and tasteless, yet highly toxic gas. The CO emission results from incomplete combustion and is prone to be the hazard of carbon monoxide poisoning.

The emission indices of NO_x and CO from the flame at $1.0<\Phi<2.0$ with fixed *Re*=8000 and at $2000<Re<10000$ with fixed Φ=1.5 are shown in Figures 5.15 and 5.16, respectively. It is found that NO_x is mainly comprised of NO, except at Φ=1.0, *Re*=8000. With fixed *Re*, both $EINO_x$ and EINO increase with an increase of Φ from 1.0 to 1.4, which is associated with the increasing temperature in the large internal recirculation zone as the flame changes from lean combustion to stoichiometric combustion in this zone. With further increasing Φ from 1.4 up to 2.0, there are slight reductions in $EINO_x$ and EINO, also following the small temperature drop in Zone 1. With fixed Φ, both $EINO_x$ and EINO increase with increasing *Re* from 2000 to 10000, which agrees with the increasing temperature in Zone 1. Thus, it is clear that thermal NO mechanism is controlling the overall NO_x emission, due to the presence of the large-size, high-temperature Zone 1 in the swirling IDF. Additionally, it is found that with the case of either fixed *Re* or fixed Φ, the variation of NO_2 emission changes in the opposite directions to NO/NO_x emissions, since NO is oxidized to NO_2 under low-temperature conditions (Syred *et al.* 1971).

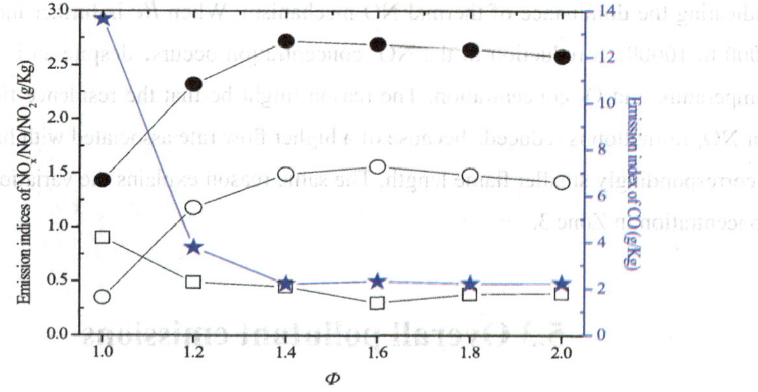

Figure 5.15 Emission indices at *Re*=8000. ●: $EINO_x$; ○: EINO; □: $EINO_2$; ★: EICO.

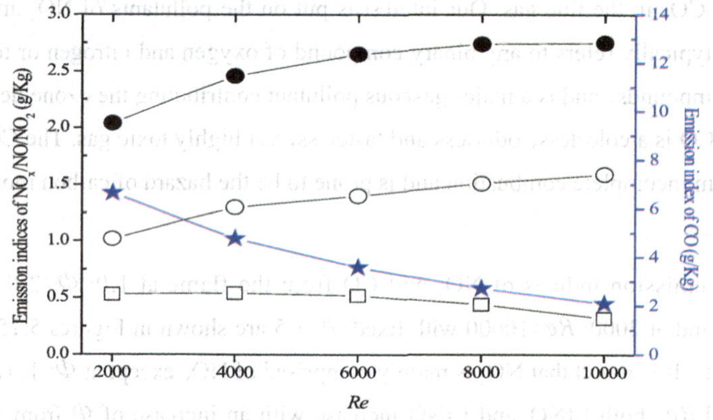

Figure 5.16 Emission indices at Φ=1.5. ●: $EINO_x$; ○: EINO; □: $EINO_2$; ★: EICO.

At *Re*=8000, EICO drops quickly from 13.64 g/kg at Φ=1.0 to 2.19 g/kg at Φ=1.4 and changes very little at higher Φ. As shown in Figure 4.4, at Φ=1.0 and *Re*=8000, the flame is very short with Zone 3 being negligibly small. The CO formed in the flame cannot be effectively oxidized in such a small Zone 3, resulting in a high value of EICO. As Φ increases to 1.2, Figure 4.4 shows that a double flame structure, which consists of a lower premixed flame involving Zone 1 and Zone 2 and an upper diffusion flame including Zone 3, becomes more discernible and meanwhile Zone 3 becomes longer. Though

CO is produced in the lower premixed flame at a higher rate, further oxidation of the CO formed by the upper diffusion flame is fast and hence a lower EICO is observed. Further drops in EICO occur until the complete combustion condition is established in Zone 1 at Φ=1.4. When the fuel-rich combustion condition has set up in Zone 1 and as Φ increases beyond 1.4, the higher CO production rate in Zone 1 is probably balanced by the higher oxidation rate in Zone 3, leading to a nearly constant EICO. At Φ=1.5, EICO decreases monotonically as *Re* increases from 2000 to 10000. The reason is that the higher air/fuel jet velocity and turbulence associated with higher *Re* enhance combustion, leading to a reduction in the amount of CO formed in the flame, as indicated by Figure 5.12. Moreover, the increasing turbulence enhances entrainment of ambient air to help further oxidation of CO in the post-combustion region.

CHAPTER 6 FLAME IMPINGEMENT HEAT TRANSFER

6.1 Introduction

Impinging flame jets have been widely used in industrial and domestic heating applications for their enhanced convective heat transfer rates. The thermal performance of a flame impingement system is dependent significantly on four factors: burner style, flame jet properties, impingement surface condition, and configuration between burner nozzle and surface.

The characteristics of the swirling IDF as an impinging flame were investigated by using a heat flux sensor to measure the radial heat flux distribution. The details of the experimental set-up have been covered in Chapter 3. Before measurements, the circulation of cooling water at a constant temperature 38℃ was set up and the amplifier was allowed sufficient time to warm up to prevent drifing during measurements. To increase the precision of measurements, the RTS and HFS are zeroed at the water temperature, 38℃ . All the radial heat fluxes were measured along a radial distance of 180 mm, starting from the stagnation point, at 4 mm interval for the first 40 mm distance, 6 mm for the next 60 mm distance and 8 mm for the last 80 mm distance. Each heat flux value is registered only after a steady state has been achieved. Based on the radial heat flux distribution, the area-integrated heat flux or the overall heat transfer rate and the heat transfer efficiency can be calculated from Equation 6.1 and Equation 6.2, respectively:

$$\dot{q}=\iint^{A}\dot{q}_{\text{local}}\mathrm{d}A=2\pi\int_{0}^{R}\dot{q}_{\text{local}}r\mathrm{d}r \tag{6.1}$$

$$\eta(\%)=\frac{\text{Output}}{\text{Input}}\times 100\%=\frac{\dot{q}}{\dot{Q}_{\text{fuel}}\rho_{\text{fuel}}\text{LHV}_{\text{fuel}}}\times 100\% \tag{6.2}$$

The area-integrated heat flux is obtained by integrating the radial distribution of the local heat flux within a circular zone of 180 mm radius centered at the stagnation point. The radial distance of 180 mm is chosen because in most cases, the local heat flux would have decayed by more than 90% beyond R=180 mm.

The carefully arranged experiments enable the individual investigation of the effects of different operational parameters including Re, Φ, H and S' on the heat transfer characteristics of the swirling IDF impingement system, in terms of the radial heat flux distribution on the target surface. The experimental tests are firstly carried out to identify the optimum nozzle-to-plate distance $H_{optimum}$ at Re=8000, Φ=1.4 and S' =9.12. Secondly, the variation of the overall equivalence ratio from 1.0 to 2.0 is performed at Re=8000, $H=H_{optimum}$ and S' =9.12. Thirdly, the Reynolds number is varied from 6000 to 10000 at Φ=1.5, $H=H_{optimum}$ and S' =9.12. Fourthly, the variation of the swirl number from 4.56 to 9.12 is implemented at Re=8000, Φ=1.5 and $H=H_{optimum}$.

6.2 Impinging flame structure

The flame image and structure of the swirling IDF in an open environment at Re=8000, Φ=1.4 and S'=9.12 are shown in Figure 6.1. When this swirling IDF impinges onto the target surface, the development of the flame along the vertical direction is impeded and the target surface forces the flame to spread outwards radially. Thus the impinging swirling IDF spreads outwards radially at a faster rate compared to that of the open swirling IDF. This higher spreading rate results in a higher level of entrainment of ambient air, which in turn results in an additional expansion of the vertical flame before its impingement on the target surface. The additional expansion refers to the enlarged cross-sectional area of a particular portion of the flame, i.e. the vertical flame below the target surface. After impingement, the flame extends outwards radially along the target surface, forming the horizontal wall jet flame. For information only, traditional impinging flame jets without induced swirl show no significant change in the cross-sectional area of the vertical flame.

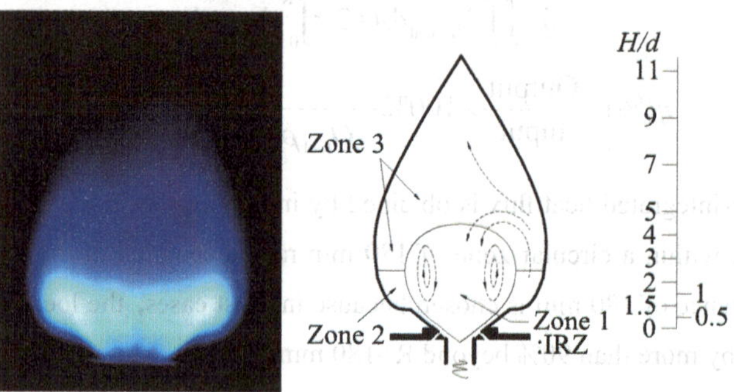

Figure 6.1 Flame image and structure of open swirling IDF at Re=8000, Φ=1.4 and S'=9.12.

Within the range of S' from 4.56 to 9.12, both the open and impinging swirling IDFs have slight change in flame appearance, such as flame shape, size and color. Additionally, the flame structure identified is also similar. Therefore, only the case of S' =9.12 is enclosed in this thesis. Basically, there are three types of flame structure for the impinging swirling IDF if the nozzle-to-plate distance is smaller than the flame height of the corresponding open swirling IDF, as shown in Figures 6.2 and 6.3. In Figure 6.3, each sketch of the impinging swirling IDF is overlapped by the contour of the corresponding open swirling IDF, designated by dashed lines. From this direct comparison, the additional expansion of the vertical flame before impingement is clearly seen in Type I and Type II, and the additional expansion in the flame almost disappears in Type III.

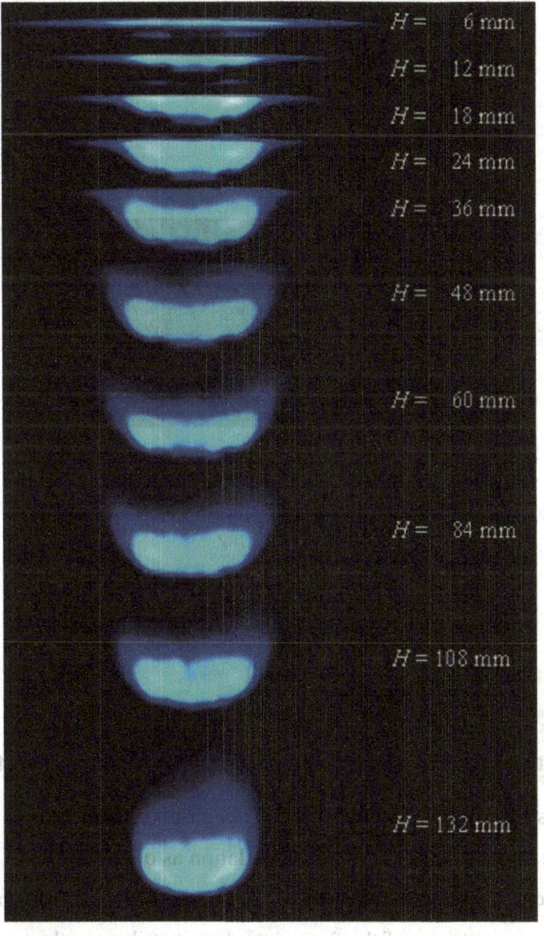

Figure 6.2 Images of impinging swirling IDF at *Re*=8000, *Φ*=1.4 and *S'*=9.12.

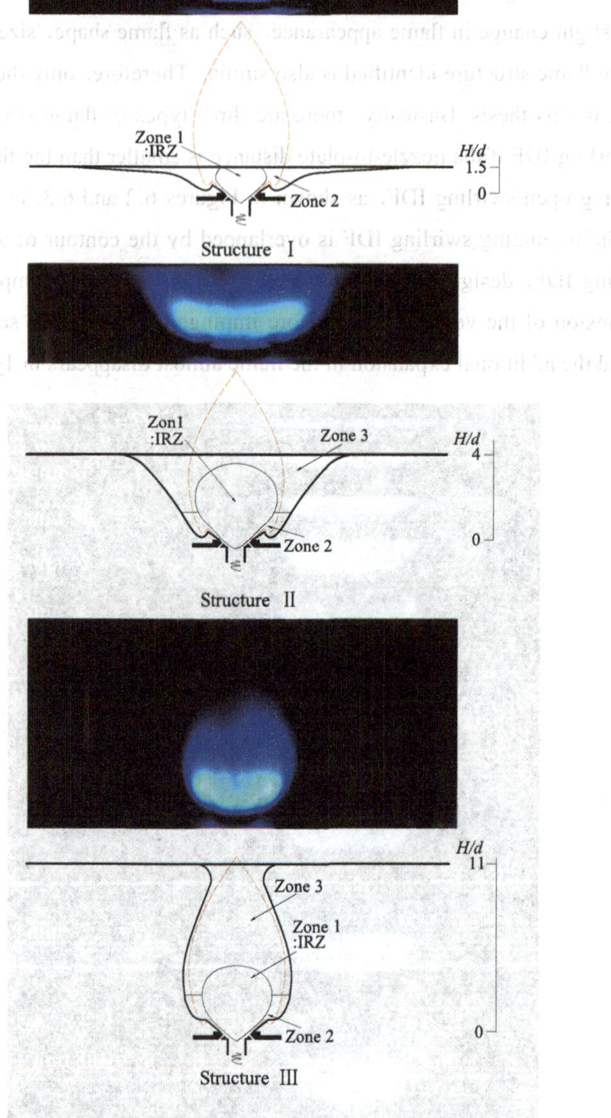

Figure 6.3 Structures of impinging swirling IDF at Re=8000, Φ=1.4 and S'=9.12.

Type I exits for $H/d \leq 2$ in which both Zones 1 and 2 impinge on the target surface, with a very small IRZ in the center. The recirculation as observed in the open swirling IDF is strongly suppressed by the target surface. The small IRZ is formed by the inward flow diverged from the mainstream of the flame jet when it strikes on the target surface at a radial position near to the stagnation point. Type II arises for $2<H/d \leq 9$ in which both Zones

1 and 3 impinge on the target surface, with a large IRZ in the center. The recirculation as observed in the open swirling IDF is slightly suppressed by the target surface. In both Type I and Type II flame structures, a portion of those fluid particles in the flame boundary which otherwise recirculate in the case of the open swirling IDF diverge outwards after colliding with the solid surface of the copper plate, enforcing more chemical reactions to occur in the wall jet region, rather than in the IRZ of the open swirling IDF. Type III appears at a sufficiently large nozzle-to-plate distance, for instance H/d=11, where the flame below the target surface resembles the open swirling IDF in terms of flame structure.

6.3 Wall static pressure

The wall static pressure exerted by a flame jet on the target surface is closely correlated with the hydrodynamic characteristics of the flame jet (Dong *et al.*2007). In this study, the radial distribution of the wall static pressure exerted by the impinging swirling IDF, as measured by a micro-manometer, is shown in Figure 6.4. The wall static pressure is proportional to the flow velocity normal to the impingement plate. Thus an increase in *Re* would lead to an increase in the wall static pressure.

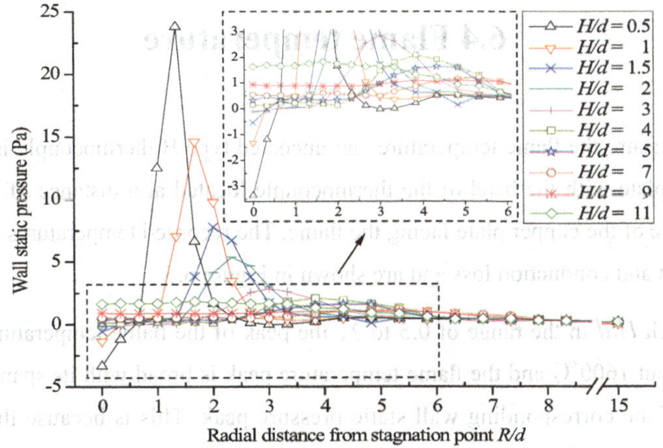

Figure 6.4 Radial profiles of wall static pressure at different *H*, *Re*=8000, *Φ*=1.4 and *S′*=9.12.

Close to the stagnation point, the wall static pressure has negative values at 0.5 ≤ *H*/

$d \leq 3$, indicating the presence of a reverse flow detaching away from the target surface. The stagnation point static pressure increases monotonically with increasing H and attains positive values at H/d=4 and above, indicating that the amount of flow being attached to the stagnation point is increasing as the target surface is elevated from the burner nozzle.

At each H/d in the range of 0.5 to 2 where Type I flame structure exits and along the radial direction, the wall static pressure increases sharply to a peak value at a radial position corresponding to the vertical flame boundary where large quantities of fluid particles in Zone 2 impinge on the target surface and then decreases rapidly to nearly atmospheric pressure in the wall jet region. At each H/d in the range of 3 to 9 where Type II flame structure arises, the wall static pressure attains peak values in the flame boundary again, and decays on both sides. It is also shown that as H/d increases from 0.5 to 9, the peak value of the wall static pressure decreases monotonically with its position shifting farther away from the stagnation point. This is because the mixing between the swirling jet and ambient air causes velocity decay and expansion of the flame jet. The mixing also results in a thicker flame boundary at higher H, as indicated by the widening span of the wall static pressure peak. At H/d=11 where Type III flame structure appears, the wall static pressure displays no clear peak and just drops along the radial direction, indicating that post-combustion products converge and strike on the target surface uniformly.

6.4 Flame temperature

To measure the flame temperature, an uncoated type B thermocouple is mounted in the copper plate with the bead of the thermocouple located at a distance of 3 mm below the front side of the copper plate facing the flame. The recorded temperatures are corrected for radiation and conduction loss and are shown in Figure 6.5.

At each H/d in the range of 0.5 to 2, the peak of the flame temperature has a high value of about 1600℃ and the flame temperature peak is broad with its span much wider than that of the corresponding wall static pressure peak. This is because that the radial position of the peak flame temperature is slightly offset to the right side of the corresponding radial position of peak wall static pressure, namely, slightly farther away from the stagnation point. Both fast air/fuel mixing and intense chemical reactions take place in the vertical flame boundary such that the wall static pressure attains its peak value at the location where the flame boundary directly impinges on the target surface. After impingement, the majority of the fluid particles diverge outwards. At short nozzle-to-plate distances, the outgoing fluid pushes some unburned fuel away from the stagnation point for further reactions. As a result, combustion is more intense slightly downstream of the flame boundary, generating the peak flame temperature at the right side of the pressure peak. Note that at H/d=0.5, the flame temperature drops heavily to 982℃ at R/d=1.3 or R=16 mm, which is a position in the closest vicinity of the burner rim and where the wall static pressure is the highest. The decrease in the flame temperature is due to the quenching effect exerted by the burner rim.

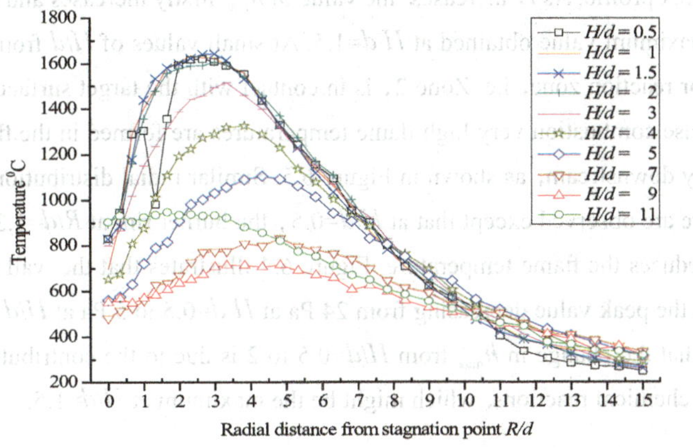

Figure 6.5 Radial profiles of flame temperature at different *H*, *Re*=8000, *Φ*=1.4 and *S'*=9.12.

As H/d increases to 3 and further to 9, the peak flame temperature drops monotonically and its radial position shifts gradually from the right to the left side of the corresponding position of peak wall static pressure. The reason is that there is less unburned fuel being diverged outwards radially after impingement and thus complete combustion eventually occurs in the vertical flame boundary before impingement on the target surface. The decrease in the peak flame temperature is due to the mixing between the swirling jet and ambient air leading to dilution and cooling down of the combustion products.

At H/d=11 and in a broad region around the stagnation point, the flame temperature becomes higher than that at H/d=9 and keeps rather uniform with no clear peak. The increase in the flame temperature is the result of the alleviated dilution associated with the disappearance of the additional expansion in the flame, thus there is less dilution and cooling-down of the combustion products.

6.5 Effects of *H* on the local heat flux

The effect of H on the radial heat flux distribution investigated at Re=8000, Φ=1.4 and S' =9.12 is shown in Figure 6.6. The profiles in Figure 6.6 can be classified to three groups. The first group corresponds to H/d=0.5 – 2, the second group pertains to H/d=3 – 9 and the third group is for H/d=11. As for the first group, Figure 6.6 shows that, from the stagnation point to R/d=15, it has double peaks at H/d=0.5 and only single peak for other values of H/d. For clarity, the denotation h_{max} is used to designate the peak value along each heat flux profile. As H increases, the value of h_{max} firstly increases and then decreases with the maximum value obtained at H/d=1.5. At small values of H/d from 0.5 to 2, the bright-color reaction zone, i.e. Zone 2, is in contact with the target surface, and because of the intense combustion very high flame temperatures are formed in the flame boundary and slightly downstream, as shown in Figure 6.5. Similar radial distributions of the flame temperature are observed except that at H/d=0.5, the burner rim at R/d=1.3 acts as a heat sink and reduces the flame temperature. Figure 6.4 illustrates that the wall static pressure drops with the peak value decreasing from 24 Pa at H/d=0.5 to 5 Pa at H/d=2. Therefore, it is clear that the change in h_{max} from H/d=0.5 to 2 is due to the contribution of heat release from chemical reactions, which might be the maximum at H/d=1.5.

With regard to the second group, Figure 6.6 illustrates that the value of h_{max} decreases monotonously from H/d=3 to 9, while the position of h_{max} shifts gradually farther away from the stagnation point. When H/d exceeds 2, a sky-blue flame, i.e. Zone 3, emerges on the top of the bright-color reaction zone and gets in contact with the target surface. Furthermore, complete combustion tends to occur in the vertical flame boundary at H/d>2. Thus the influence of heat release on the local heat flux could be ignored and the decrease in h_{max} from H/d=3 to 9 is attributed to the decay of impinging velocity and flame temperature, both being incurred by the mixing between the swirling jet and ambient air. As for the last group, i.e., at H/d=11, Figure 6.6 shows that in a broad region around the stagnation point, the local heat flux becomes substantially higher than that at H/d=9, and remains uniform at R/d<3.3. A transition in flame structure from Type II to Type III occurs as H/d increases from 9 to 11. At H/d=11, the nozzle-to-plate distance is comparable to the flame height of the corresponding open swirling IDF, and the impinging IDF behaves similarly to the open IDF in that most of the fluid particles in Zone 3 converge towards the burner axis before impingement. Thus the additional expansion in the vertical flame disappears and upon impingement the fluid particles, which are mainly post-combustion products, are diverged outwards radially by the target surface. Because of the disappearance of the additional expansion, mixing between the swirling jet and ambient cold air is reduced, and both flame temperature and impinging velocity are recovered to a certain extent, bringing increases to the local heat flux. In the far wall jet region, as shown in Figure 6.6, there is little difference among the radial heat flux profiles, illustrating that combustion has reached completion and it is the combustion products which form a hot gas layer that heats up the target surface.

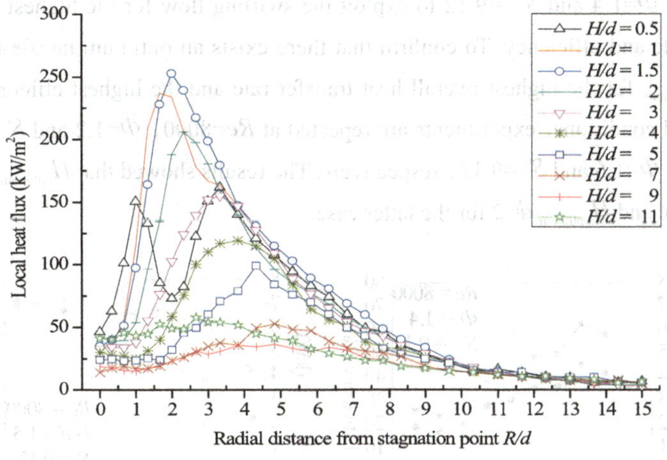

Figure 6.6 Radial profiles of heat flux at different H, Re=8000, Φ=1.4 and S'=9.12.

In conclusion, the swirling effect influences heat transfer in three ways. Firstly, the heat transfer at the stagnation point is severely deteriorated by swirl. Secondly, both high impinging velocity and high flame temperature dwell in the flame boundary, and the local heat flux is the highest at a certain radial position, following the position of either the peak impinging velocity or the peak flame temperature. Thirdly, as H increases, the swirling jet brings the position of h_{max} gradually farther away from the stagnation point. For a non-swirling IDF, Sze *et al.* (2004) reported that the parameter of H influences the radial heat flux profiles in two ways. Firstly, at small H, the heat flux increases radially until a peak value is reached and then drops steadily in the radial direction. The low heat flux at the stagnation point is due to the existence of a cool core. At higher H, the heat flux is higher and the position of peak heat flux is farther away from the stagnation point. Secondly, at sufficiently high H, the peak heat flux occurs at the stagnation point due to the disappearance of the cool core, and the heat flux decreases radially.

Based on Equations 6.1 and 6.2, the area-integrated heat flux or the overall heat transfer rate and the heat transfer efficiency of different H at Re=8000, Φ=1.4 and S' =9.12 are calculated and shown in Figure 6.7 (a) .It is seen that the overall heat transfer rate $\dot{q}$ increases from 3.64 KW at H/d=0.5 to 4.27 KW at H/d=1.5, decreases to 1.72 KW at H/d=9 and increases slightly to 1.89 KW at H/d=11. Since both Re and Φ are fixed, the heat transfer efficiency and the overall heat transfer rate have the same trend of variation with H. Therefore, both the maximum $\dot{q}$ and the maximum η occur at H/d=1.5, which means that this is the optimum nozzle-to-plate distance for the operational condition of Re=8000, Φ=1.4 and S' =9.12 to exploit the swirling flow for the highest overall heat transfer rate and efficiency. To confirm that there exists an optimum nozzle-to-plate distance $H_{optimum}$ for the highest overall heat transfer rate and the highest efficiency at other operational conditions, experiments are repeated at Re=8000, Φ=1.2 and S' =9.12, and Re=8000, Φ=1.6 and S' =9.12, respectively. The results showed that $H_{optimum}/d$=1 for the former case and $H_{optimum}/d$=2 for the latter case.

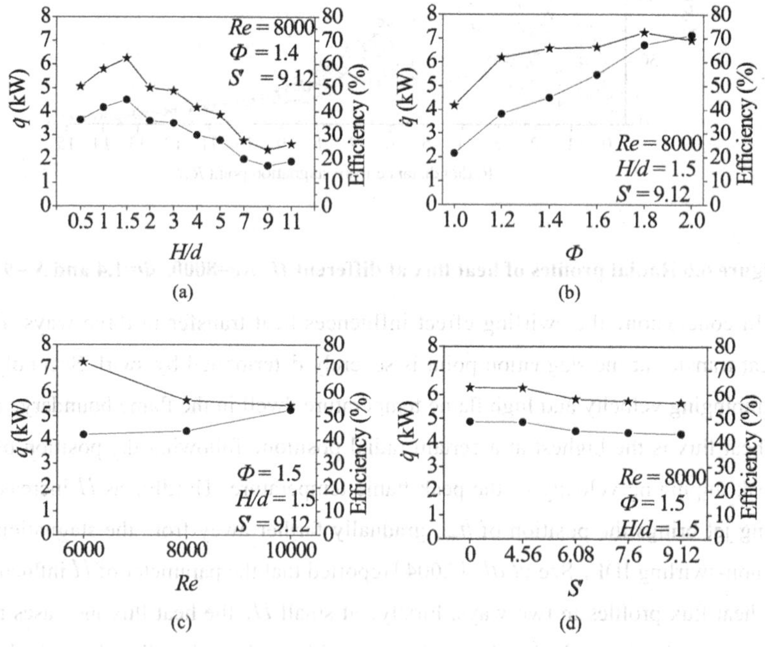

Figure 6.7 Variation of overall heat transfer rate and heat transfer efficiency with operational parameters: (a) H; (b) Φ; (c) Re; (d) S'; "•" for overall heat transfer rate and" ★ " for heat transfer efficiency.

6.6 Effects of Φ on the local heat flux

The effect of Φ on the radial heat flux distribution investigated at Re=8000, H/d=1.5, and S' =9.12 is shown in Figure 6.8. The local heat flux along each profile is very low at the stagnation point, increases steeply to a peak value in the radial direction and then drops gradually in the wall jet region. As mentioned earlier, this pattern of radial heat flux distribution is due to the swirling effect. The six profiles overlap each other in the region close to the stagnation point, depart towards reaching their different peaks, and differ very much despite following the same trend of decline in the wall jet region.

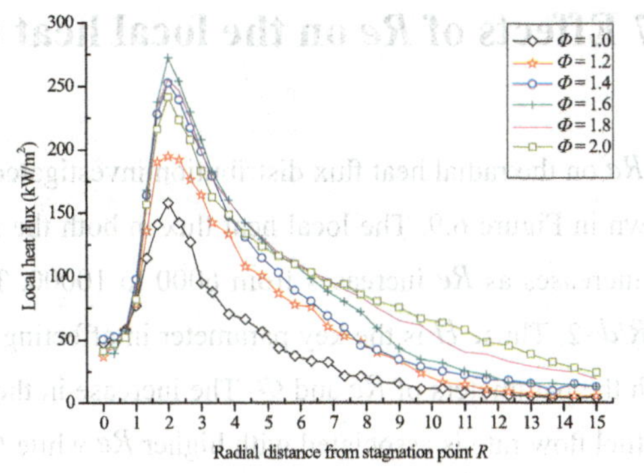

Figure 6.8 Radial profiles of heat flux at different Φ, Re=8000, H/d=1.5 and S'=9.12.

As Φ changes in the range of $1.0 \leq \Phi \leq 2.0$ and close to the stagnation point, the profiles overlap each other, showing that the stagnation point heat transfer remains severely suppressed by the swirling effect and thus the influence of Φ on the radial distribution of the local heat flux is negligibly small is this region. The profiles exhibit a large variation in the value of h_{max} which first increases and then decreases since the increasing value of Φ from 1.0 to 2.0 leads the flame to fuel-lean, stoichiometric and fuel-rich conditions, resulting in the maximum h_{max} attained at Φ=1.6. However, h_{max} occurs at nearly the same location of R/d=2, which indicates that there is little change in the hydrodynamic characteristics of the impinging flame jet. After reaching their peaks, these profiles decline steadily and the rate at which a profile declines is lower at higher Φ in the wall jet region, manifesting that when Φ increases, there is more unburned fuel pushed outwards radially to react with air in the wall jet region and therefore the flame temperature in this region tends to increase and the local heat flux becomes higher.

The overall heat transfer rate and the heat transfer efficiency of different Φ, at Re=8000, $H_{optimum}/d$=1.5 and S'=9.12 are shown in Figure 6.7（b）. An increase in Φ at fixed Re means that the fuel flow rate is increasing while the air flow rate is fixed. Therefore, the value of $\dot{q}$ increases monotonically from 2.15 KW at Φ=1.0 to 7.14 KW at Φ=2.0, due to the increasing energy input. The maximum heat transfer efficiency occurs around Φ=1.8 and drops on both ends. This result is similar to the finding of Ng *et al.*（2007), who studied in details the efficiency of the heat transfer process of a non-swirling IDF, and stated that at fixed Re and H, the heat transfer efficiency increases to a maximum value and then decreases as Φ increases.

6.7 Effects of *Re* on the local heat flux

The effect of Re on the radial heat flux distribution investigated at Φ=1.5, H/d=1.5 and S'=9.12 is shown in Figure 6.9. The local heat flux in both the stagnation region and the wall jet region increases as Re increases from 6000 to 10000. The radial position of h_{max} is invariant at R/d=2. Thus, H is the key parameter in affecting the radial position of h_{max}, compared with the parameters of Re and Φ. The increase in the local heat flux is because that a higher fuel flow rate is associated with higher Re while Φ is kept unchanged, which leads to more heat released. Further, higher Re generally induces higher impinging velocity, better mixing of the air/fuel, higher level of turbulence and thus higher flame temperature, all of which could enhance the heat transfer rates.

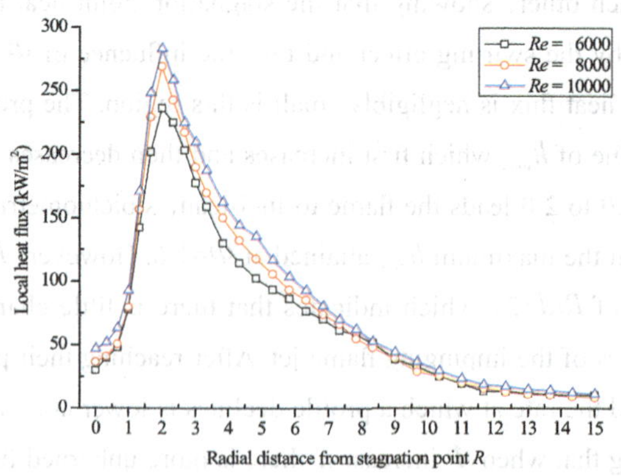

Figure 6.9 Radial profiles of heat flux at different *Re*, *Φ*=1.5, *H*/*d*=1.5 and *S'*=9.12.

The overall heat transfer rate and the heat transfer efficiency of different Re, at Φ=1.5, H/d=1.5 and S' =9.12 are shown in Figure 6.7(c).

An increase in Re at fixed Φ means that both the fuel and air flow rates are increasing, so that there is an increase in the energy input to the flame impingement system and therefore the overall heat transfer rate $\dot{q}$ increases monotonously from 4.21 kW at Re=6000 to 5.26 kW at Re=10000. Similar effect of Re on the overall heat transfer was observed by Sze *et al.* (2004) and Ng *et al.* (2007). To the contrary of $\dot{q}$, the heat transfer efficiency decreases as Re increases. This result is similar to the finding of Ng *et al.* (2007) who attributed the decreasing efficiency with Re to that more energy is transferred beyond the integration area, which might not be true in this study because of the different range of Re between the two studies. The possible reason in the current study is more likely related to the increase in the amount of entrained ambient air at higher Re. The entrained ambient cold air absorbs a significant amount of energy and takes away a larger portion of heat such that the portion to the target surface is reduced.

6.8 Effects of S' on the local heat flux

The degree of swirl is altered by adjusting the relative magnitude of tangential and axial flow rates of air for the swirl burner. For guarantee of the formation of the IRZ in the swirling IDF, the adjustment of swirl number is restricted to the high-swirl range of $4.56 \le S' \le 9.12$. The effect of S' on the radial heat flux distribution investigated at Re=8000, Φ=1.5 and $H_{optimum}/d$=1.5 is shown in Figure 6.10.

Generally, a higher swirl number tends to move the radial position of h_{max} outwards radially because of a higher jet spreading rate. In the study of Lee *et al.* (1997), the radial position of h_{max} is situated markedly further away from the stagnation point when the actual swirl number S' increases from 0.21 to 0.77. However in this study, the displacement of the radial position of h_{max} is not observed at H/d=1.5 and in the range of $4.56 \le S' \le 9.12$, probably because the divergent outlet confines the evolution of the swirling jet at short nozzle-to-plate distances. Specifically, the centrifugal force induced by such high swirl is so strong that the swirling jet follows the curvature of the divergent outlet. In other words, the angle of spread of the flame is always 450 which is half of the cone angle of the divergent outlet. As shown in Figure 10, the radial position of h_{max} is maintained at R/d=2, and the value of h_{max} increases monotonically as S' increases from 4.56 to 9.12, which might be due to the higher flame temperature associated with better air/fuel mixing and higher level of turbulence. Except the difference in the value of h_{max}, the other parts of the profiles are close to each other, with the local heat flux in the wall jet region being slightly lower at higher S'.

As S' increases, there is a slight decrease in the overall heat transfer rate, as shown in Figure 6.7 (d), decreasing monotonically from 4.85 kW to 4.36 kW as S' increases from 4.56 to 9.12. Thus there is no enhancement in heat transfer caused by the elevated h_{max} and actually the overall heat transfer rate decreases with S', which shows a negative effect of swirl on heat transfer. Figure 6.7 (d) also shows that the heat transfer efficiency decreases with increasing S', which is evidence that the swirl has an unfavorable effect on heat transfer.

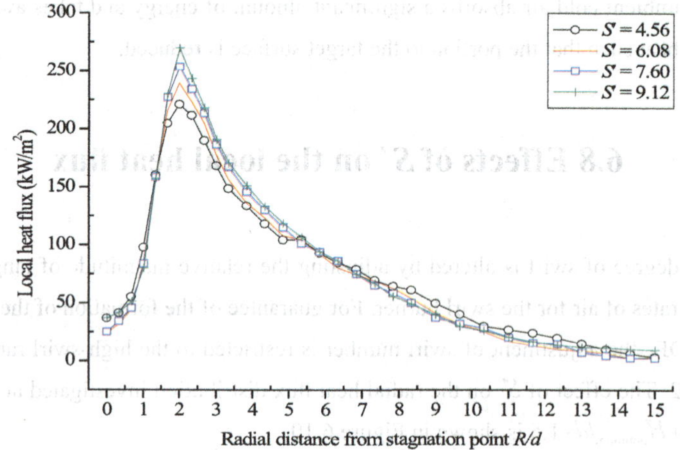

Figure 6.10 Radial profiles of heat flux at different *S'*, *Re*=8000, *Φ*=1.5, and *H*/*d*=1.5.

CHAPTER 7 COMPARISON OF FLAMES

7.1 Introduction

The characteristics of the swirling IDF including flame appearance, structure, temperature, emission and impingement heat transfer have been analyzed in details in Chapters 4, 5 and 6. It is always of interest to compare different flames to better understand their particular behaviors. Hence, experiments were carried out to compare IDFs with and without swirl under the same or similar experimental conditions. Flame appearance, overall pollutant emissions and impinging heat transfer characteristics of the two IDFs are compared. Then, the comparison of the swirling IDF with a swirling premixed flame, in terms of flame appearance, flame temperature, in-flame gaseous species and overall pollutant emissions, are conducted.

7.2 Comparison of IDFs with and without swirl

The swirl burner together with a non-swirl burner, co-flowing fuel and air jets is shown in Figure 3.4. The diameters of both air/fuel ports and the center-to-center spacing between the ports are the same for the two burners, so direct comparison can be made between the swirling and non-swirling IDFs which are made to operate under the same operational conditions.

7.2.1 Comparison of flame appearance

The flame images and lengths of the swirling and non-swirling IDFs at fixed Φ=1.5 are shown in Figure 7.1. The appearances of the IDFs with and without induced swirl are observed to be distinctively different, despite their similar burner geometry identical air/fuel flow rates and the same air jet Reynolds number. Basically the non-swirling IDFs are more slender while the swirling IDFs are larger in cross-sectional diameter. For the non-swirling IDFs, Figure 7.1 (He upper) shows an increase of the flame length from Re=2000 to Re=6000 and then the (He lower) shows flame length remains fairly constant at higher Re. The flame consists of a premixed bluish flame close to the burner exit and a downstream diffusional yellow flame. The flame becomes less diffusional in nature with an increase in *Re*. In contrast, for the swirling IDFs, the flame length decreases from 173 to 105 mm with an increase in *Re* from 2000 to 10000. The swirling IDF becomes mainly premixed in nature when *Re* is equal to and larger than 6000. This is because more intensive mixing and hence more complete combustion occur in the IRZ of the swirling IDF. It is also observed that with increasing *Re*, the IRZ grows in size with the vortex centers shifting slightly away from the nozzle axis and moving downstream.

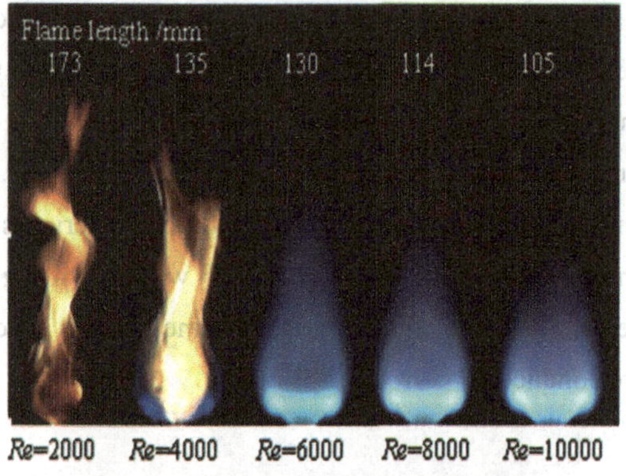

Figure 7.2 Flame images of IDFs with (upper) and without (lower) induced swirl at Φ=1.5.

The effect of Φ on the flame appearance at fixed *Re*=8000 is shown in Figure 7.2. At fixed *Re*, an increase in Φ corresponds to an increase in the amount of supplied fuel.

Figure 7.2 shows an increase in the flame length for both the swirling and the non-swirling IDFs, while the non-swirling IDFs are much longer than the swirling IDFs. Again the non-swirling IDFs are observed to consist of a premixed bluish flame overlapped by a diffusion flame, with a growth in the downstream diffusion flame as Φ increases from 1.0 to 2.0. However, the swirling IDF remains fairly premixed in nature with the post-combustion zone, which is weakly diffusional in nature, becoming more and more observable with increasing Φ.

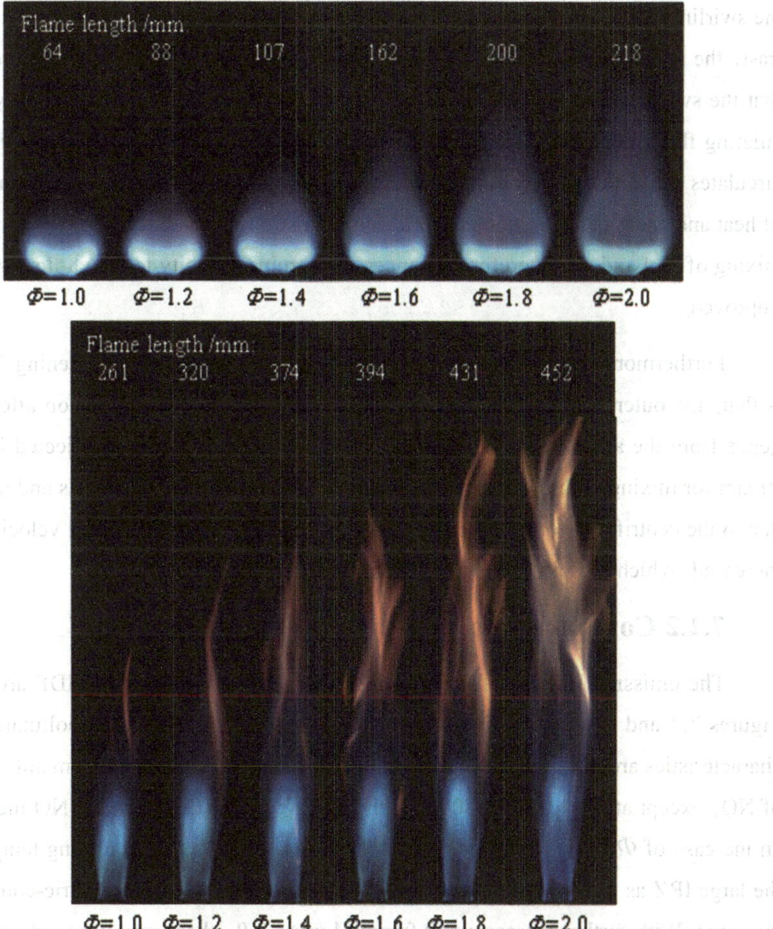

Figure 7.1 Flame images of IDFs with (upper) and without (lower) induced swirl at *Re*=8000.

It concluded that the flame length of swirling IDFs is dependent on both *Re* and Φ, just as that in premixed flames (Kwok *et al.*2005). At fixed *Re*, both swirling and non-swirling IDFs lengthen with increasing Φ due to increased fuel supply. At fixed Φ, swirling IDFs shorten with increasing Re because of higher air/fuel jet velocity and turbulence level which improve air/fuel mixing and increase the decay of axial velocity component. The effect of Re on the flame length of non-swirling IDFs is similar to that of a normal diffusion flame: once a fully turbulent flame is formed, further increase in Re has only little effect on the flame length.

The flame stability is also different for the IDFs with and without induced swirl. For the swirling IDF, a highly stable flame is observed for Re up to 20000 at Φ=1.5. In contrast, the non-swirling IDF is stable for Re up to 14000 only at Φ=1.5. So we conclude that the swirling IDF is more stable because the flame stability is enhanced by the recirculating flow induced by swirl. The IRZ plays a powerful role in flame stabilization. It circulates heat and active chemical species to the root of the flame. It also acts as a storage of heat and chain carriers, and is associated with strong turbulence which is helpful in fast mixing of fuel and air. Therefore, chances for burning velocity to match flow velocity are improved.

Furthermore, the IRZ plays an important role in flame length shortening. The reason is that: the outer boundary of the swirling jet flow expands radially soon after its emergence from the air port; at the position of initial expansion, fuel is injected into the air stream for mixing with the air; particles of the fuel and air eject outwards and tangentially due to the centrifugal force caused by swirl; as a result, the rate of axial velocity decay is increased, which is a primary parameter influencing flame length.

7.2.2 Comparison of emission index

The emission indices of the swirling IDF and the non-swirling IDF are shown in Figures 7.3 and 7.4, respectively, which illustrate that their overall pollutant emission characteristics are different. For the swirling IDF, it is found that NO_x is mainly comprised of NO, except at Φ=1.0, Re=8000. With fixed Re, both $EINO_x$ and EINO increase with an increase of Φ from 1.0 to 1.4, which is associated with the increasing temperature in the large IRZ as the flame changes from lean-combustion to stoichiometric-combustion in this zone. With further increasing Φ from 1.4 up to 2.0, there are slight reductions in $EINO_x$ and EINO, also following the small temperature drop in Zone 1. With fixed Φ, both $EINO_x$ and EINO increase with an increasing Re from 2000 to 10000, which also agrees with the increasing temperature in Zone 1. This is because the thermal NO mechanism is controlling the overall NO_x emission, due to the presence of the large-size, high-temperature Zone 1 in the swirling IDF.

At *Re*=8000, EICO drops quickly from 13.64 g/kg at Φ=1.0 to 2.19 g/kg at Φ=1.4 and changes very little at higher Φ. The EICO at Φ=1.0 is even higher than that of the non-swirling IDF operating under the same condition. As Φ increases to 1.2, though CO is produced in the lower premixed flame at a higher rate, further oxidation of the CO formed by the upper diffusion flame is fast and hence a lower EICO is observed. Further drops in EICO occur until the complete combustion condition is established in Zone 1 at Φ=1.4. When the fuel-rich combustion condition has set up in Zone 1 and Φ increases beyond 1.4, the higher CO production rate in Zone 1 is probably balanced by the higher oxidation rate in Zone 3, leading to a nearly constant EICO.

At Φ=1.5, EICO decreases monotonically as *Re* increases from 2000 to 10000. The reason is that the higher air/fuel jet velocity and turbulence associated with higher *Re* enhance combustion, leading to a reduction in the amount of CO formed in the flame. Moreover, the increasing turbulence enhances entrainment of ambient air to help further oxidation of CO in the post-combustion region.

Figure 7.4 shows that NO_x of the non-swirling IDF is mainly composed of NO except at Φ=1.5, *Re*=10000. The non-swirling IDF manifests a slight drop in $EINO_x$ and EINO with an increase of Φ from 1.0 to 2.0, at fixed *Re*=8000 and a slight rise in both $EINO_x$ and $EINO_2$ with an increase of *Re* from 2000 to 8000 at fixed Φ=1.5. In either case, the conversion of NO to NO_2 makes EINO change in the opposite direction to NO_2 emission. Mostly, $EINO_x$ is lower in the non-swirling IDF than that in the swirling IDF except at Φ=1.0~1.2, *Re*=8000 and at *Re*=2000, Φ=1.5. Two reasons are likely to account for the higher $EINO_x$ in the swirling IDF. One is that a higher temperature is expected in the swirling IDF due to better combustion. The other reason is that the swirling IDF might have a residence time of an equivalent level to that in the non-swirling IDF, based on the assumption that under the same conditions, the decrease in residence time caused by shorter flame length in the swirling IDF might be offset by the increase in residence time caused by the recirculating flow.

EICO of the non-swirling IDF is quite stable in the range of $1.0<\Phi<2.0$ at *Re*=8000, indicating a balance between the CO production and CO oxidation. EICO is higher than the corresponding swirling IDF at $\Phi \geq 1.2$, which means that the combustion is better in the swirling IDF at higher level of Φ due to the stronger air/fuel mixing in the IRZ. In the range of 2000<*Re*<10000 at Φ=1.5, EICO is higher than that of the swirling

IDF but has the same trend of variation, which again indicates that the combustion is more intense and more complete in the swirling IDF.

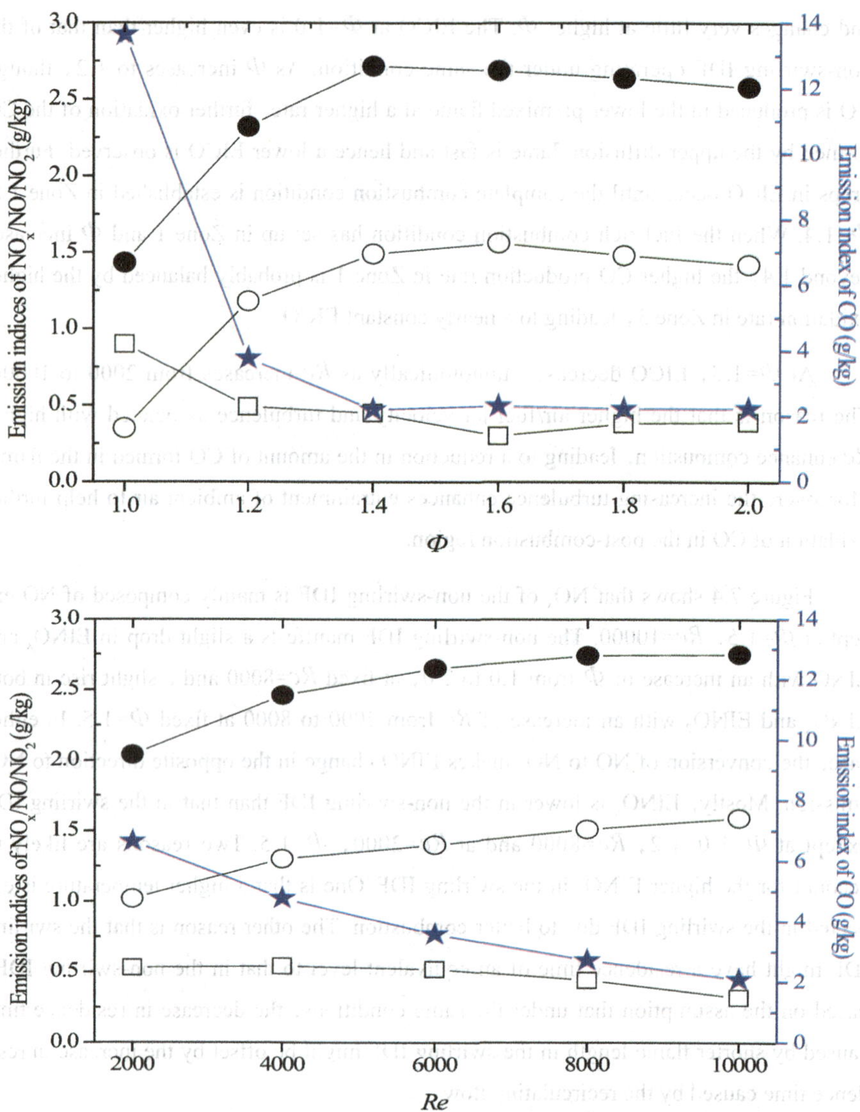

Figure 7.3 Emission indices (g/kg) at *Re*=8000 (upper) and at *Φ*=1.5 (lower) for IDF with swirl. ●: $EINO_x$; ○: EINO; □: $EINO_2$; ★: EICO.

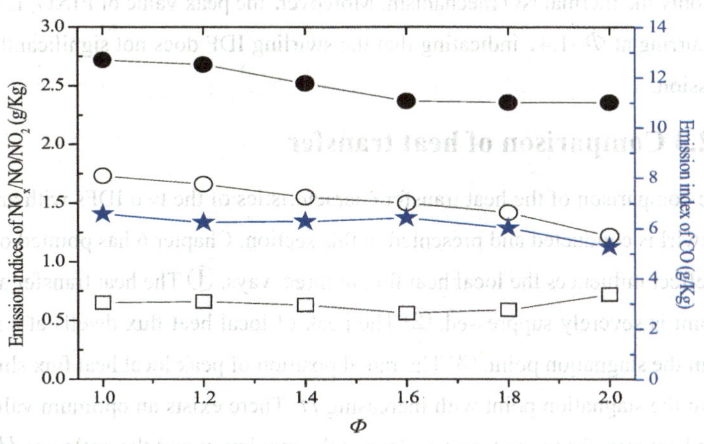

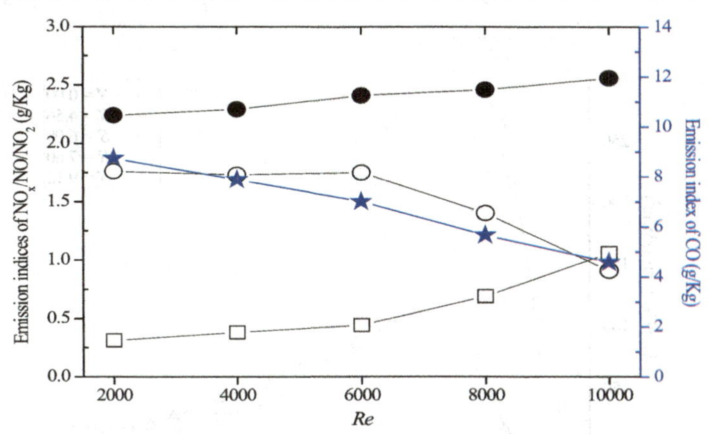

Figure 7.4 Emission indices (g/kg) at *Re*=8000 (upper) and at *Φ*=1.5 (lower) for IDF without swirl. ●: $EINO_x$; ○: EINO; □: $EINO_2$; ★: EICO.

$EINO_x$ obtained from the IDFs in the present study is used for comparison of NO_x emissions from other studies. The IDF discussed in the work of Sze *et al.* (2006), consisting of an inner air jet and twelve outer fuel jets, has an $EINO_x$ of 2.75 g/kg at *Φ*=1.0, *Re*=2500, which increases to a peak value of 3.2 g/kg at *Φ*=1.2, decays exponentially to below 2 g/kg at *Φ*=3.0 and finally decreases gradually to 1.75 g/kg at *Φ*=6.0 while *Re* is fixed at 2500. They attributed the bell-shaped distribution of $EINO_x$ to the thermal NO mechanism and NO reburn mechanism. In this study, the swirling flame presents a similar bell-shaped distribution of $EINO_x$ as Φ varies from 1.0 to 2.0 at fixed Re=8000 but clarified with only the thermal NO mechanism. Moreover, the peak value of $EINO_x$ is about 2.72 g/kg, occurring at Φ=1.4, indicating that the swirling IDF does not significantly increase NO_x emission.

7.2.3 Comparison of heat transfer

The comparison of the heat transfer characteristics of the two IDFs with and without induced swirl is conducted and presented in this section. Chapter 6 has pointed out that the swirling effect influences the local heat flux in three ways. ① The heat transfer at the stagnation point is severely suppressed. ② The peak of local heat flux dwells at a radial distance from the stagnation point. ③ The radial position of peak local heat flux shifts farther away from the stagnation point with increasing H. There exists an optimum value of H at which the heat transfer to the target surface is the maximum and the optimum H increases with increasing Φ while the Reynolds number and the swirl number are unchanged.

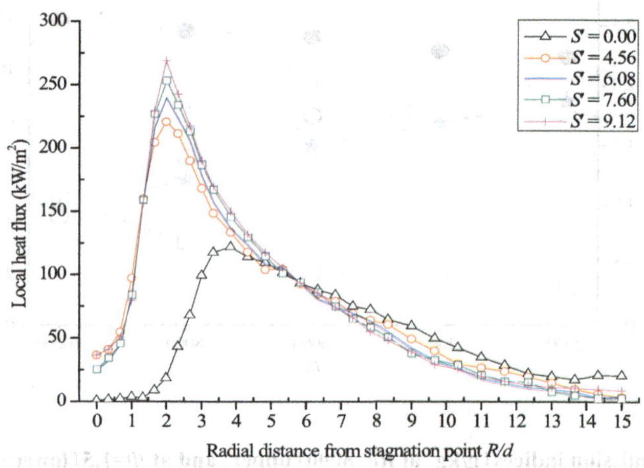

Figure 7.5 Radial profiles of local heat flux at different S', Re=8000, Φ=1.5, and H/d=1.5.

Non-swirling IDFs have been extensively investigated in literature (Dong *et al.*2007, Ng *et al.*2007 and Sze *et al.*2004).

In the present study, a non-swirling IDF with a swirl number of S'=0 exhibits similar behaviour to that of IDFs studied in the literature. At small nozzle-to-plate distances, there exists a cool core in the centre of the flame adjacent to the burner exit. The cool core mainly consists of air supplied from the central air port and leads to very low heat flux in the stagnation region. The cool core gradually disappears as complete combustion is reached at higher nozzle-to-plate distances, displaying a bell-shaped radial heat flux distribution. At even higher H, the local heat flux drops as combustion products heat up the target surface.

The radial heat flux profile at Re=8000, Φ=1.5, H/d=1.5 and S'=0 is shown in Figure 7.5. The nearly zero heat flux in the vicinity of the stagnation point is dictated by the cool core. The radial position of h_{max} is at R/d=3.8 instead of R/d=2 for the swirling IDFs and the value of h_{max} is 122 kW/m^2 which is much lower than those of the swirling IDFs. This illustrates that the induced swirl promotes rapid mixing of the air/fuel and boosts intense combustion so that the cool core present in the case of the non-swirling IDF disappears when swirl is introduced. Thus, the swirling IDFs achieve complete combustion at lower H, when compared with the non-swirling IDFs. At R/d>5.8, the local heat flux of the non-swirling IDFs become higher than that of the swirling IDFs. This shows that because of less intense mixing between the fuel and air in the non-swirling IDFs, more unburned fuel is forced to the wall jet region for further reactions with air. This particularly higher heat flux in the wall jet region makes the non-swirling IDFs possess an overall heat transfer rate higher than those of the swirling IDFs. As shown in Figure 7.6, the overall heat transfer rate for the non-swirling IDFs at S'=0 has an overall heat transfer rate of 4.86 kW which is higher than those of the swirling IDFs. The heat transfer efficiency of the non-swirling IDFs is also higher than those of the swirling IDFs, as shown in Figure 7.6.

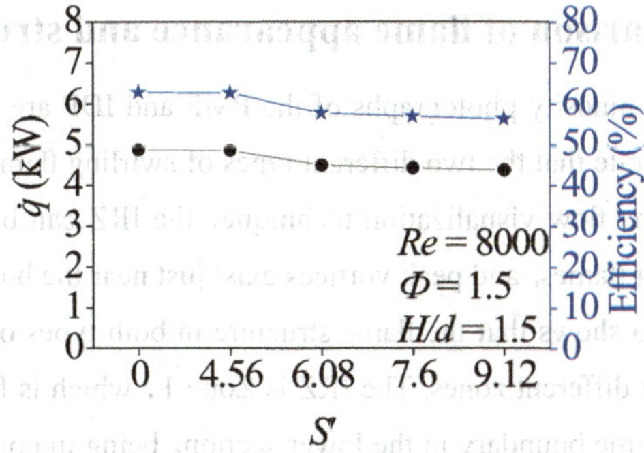

Figure 7.6 Variation of overall heat transfer rate and heat transfer efficiency with swirl number S'; "•" for overall heat transfer rate and" ★ "for heat transfer efficiency.

Ng *et al.* (2007) suggested that for achieving the highest overall heat transfer rate, the value of H should correspond to the range of complete combustion region. In this study, the non-swirling IDF operating at Re=8000 and Φ=1.4 achieves complete combustion at H/d=6 and attains the highest overall heat transfer rate of 5.73 kW, with a bell-shaped radial heat flux distribution and h_{max}=267 kW/m^2 occurring at the stagnation point. From the findings mentioned earlier, for the swirling IDF at Re=8000, Φ=1.4 and S'=9.12, $H_{optimum}/d$=1.5. Thus a comparison of the highest overall heat transfer rates of the impinging swirling and non-swirling IDFs at their individual optimum nozzle-to-plate distances is conducted and the results show that the swirl has an adverse effect on the heat transfer and reduces the overall heat transfer rate. The reduction from 5.73 kW of the non-swirling IDF at $H_{optimum}/d$=6 to 4.27 kW of the swirling IDF at H/d=1.5 is as high as 25%.

7.3 Comparison of swirling IDF with premixed flame

Two swirl-stabilized flames, pre-mixed flame (PMF) and inverse diffusion flame (IDF), leading to different mixing mechanisms, are compared under the same fuelling and airing rates. The comparison is performed in terms of flame appearance, temperature, in-flame gaseous species emissions and overall pollutant emissions. The swirl burner shown in Figure 3.3 allows formation of two different types of flame. One is PMF, generated by use of the central port for a mixture of fuel and air, with the twelve small ports sealed by plasticine. The other is IDF which uses the central port for air and the twelve small ports for fuel.

7.3.1 Comparison of flame appearance and structure

Direct flame luminosity photographs of the PMF and IDF are shown in Figures 7.7 and 7.8, which illustrate that the two different types of swirling flames have similar shape and size. Aided by the flow visualization technique, the IRZ can be clearly observed in both types of swirling flames, and peak vortices exist just near the burner rim, as shown in Figure 7.9 which also shows that the flame structure in both types of swirling flames can be described by three different zones. The IRZ is Zone 1, which is formed by the reverse flow. Zone 2 is the flame boundary in the lower section, being in contact with ambient air on the outer side and with Zone 1 on the inner side. Zone 3 is the flame boundary in the upper section, which includes the post-combustion region overlapping Zones 1 and 2.

The similar flame shape, size and structure of the PMF and IDF are dictated by the similar operational conditions. In Figure 7.7 the air flow rate is fixed at $\dot{Q}$ =67.86T/min, and the fuel flow rate increases from 2.36T/min to 4.71T/min, leading Φ to range from 1.0 to 2.0. In Figure 7.8, the air flow rate increases from 16.96T/min to 84.82T/min, and the fuel flow rate increases accordingly to keep Φ fixed at 1.5. These operational conditions mean that for each flame shown in the figures, the fuel flow rate is always smaller than 7% of the air flow rate. In the case of the PMF, due to the contribution from the fuel flow, the value of Re, calculated from the gaseous stream flow rate through the central port, is slightly larger than that of the IDF. However, the maximal difference in Re in the cases of the PMF and IDF, is less than 7%, which is negligibly small. Furthermore, based on the fact that the same swirl generator, i.e. the swirl chamber is used for the PMF and IDF, it is reasonable to assume that both Re and S', which are key parameters representative of the fluid-dynamic characteristics of the swirling jet flow effusing from the central port, do not undergo significant changes between the cases of the PMF and IDF.

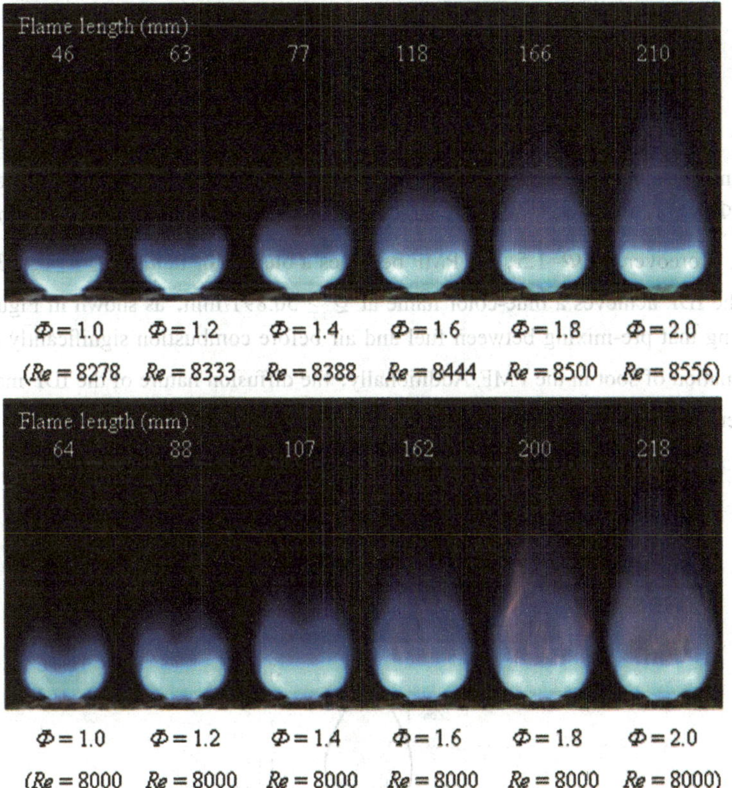

Figure 7.7 Flame images of PMF (upper) and IDF (lower) at $\dot{Q}$=67.86T/min.

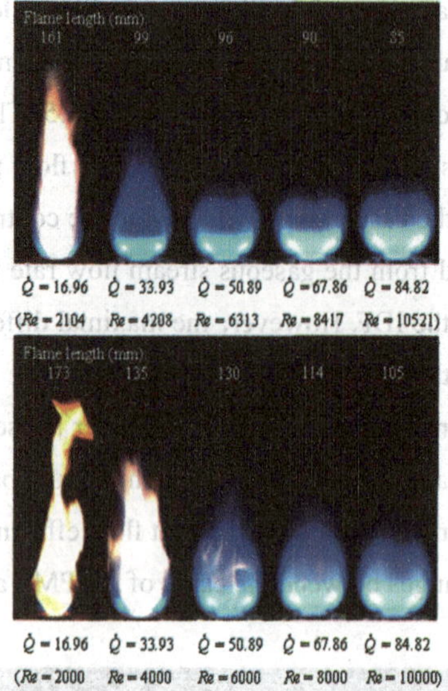

Figure 7.8 Flame images of PMF (upper) and IDF (lower) at Φ=1.5.

A close examination of the two swirling flames reveals that the IDF is more diffusional in nature. At $\dot{Q}$ =67.86T/min, yellowish flame brushes intermittently appear in the IDF at $\Phi \geq 1.8$, while there are no visible yellowish flames in the PMF, as shown in Figure 7.7. Moreover, at Φ=1.5, the PMF becomes a blue-color flame at $\dot{Q} \geq 33.93$T/min, while the IDF achieves a blue-color flame at $\dot{Q} \geq 50.89$T/min, as shown in Figure 7.8, indicating that pre-mixing between fuel and air before combustion significantly reduces the formation of soot in the PMF. Additionally, the diffusion nature of the IDF makes the flame length slightly longer than that of the PMF.

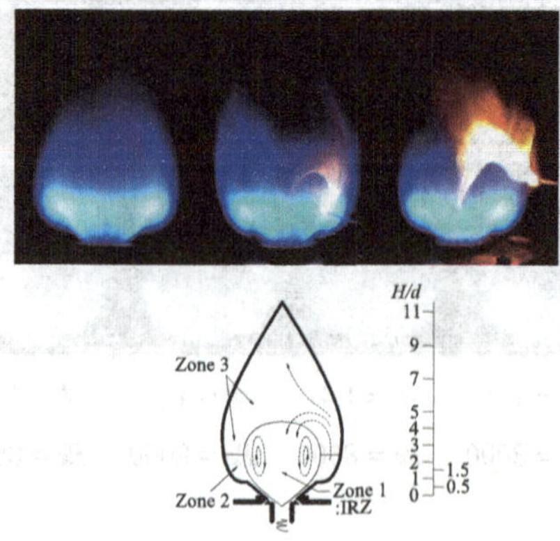

Figure 7.9 Flame structures of PMF and IDF.

7.3.2 Comparison of flame stability

In the present study, the flame stabilization mechanism is considered to be accordant with literature Syred and Beer (1974). Namely, at the edge of the IRZ where the flame front resides, the flow velocity and the flame speed are matched, aided by the recirculation of heat and active chemical species. Zone 2, an annular region that encloses the IRZ, is featured by a bright and navy-blue color, indicating that a curved annular flame front is located at the edge of the IRZ. The IRZ, being a storage of heat and chain carriers, carries heat and active chemical species towards the root of the flame, and ignites the fresh mixture of fuel and air. Therefore, between the forward going reactants and the reverse flow of the combustion products is the flame front which is associated with intensive mixing and combustion.

The flame stability of the PMF and IDF can be described by the lean blow-off limit (LBO) and LBO is determined by maintaining a constant volumetric flow rate of the jet flow through the central port and incrementally reducing the fuel flow rate until the flame becomes unstable and eventually blows off. Fig.7.10 shows a comparison of the results for the PMF and IDF plotted against the bulk velocity $U_0=Q / A_e$, where A_e is the open area of the central port. Towards LBO, either the PMF or IDF becomes visibly weaker and smaller, and eventually disappears. The LBO limit of the PMF is higher than that of the IDF and seems more sensitive to U_0 in the range of $U_0<7.5$ m/s where flashback occurs when Φ gets close to 0.9. At higher U_0, higher flow velocity in the PMF prevents the occurrence of flashback and the LBO limit shows an increasing trend with U_0. The better stability of the IDF is due to the diffusion process, i.e. the fuel diffuses into the air and causes simultaneous mixing and combustion, thus the rate of combustion depends on the rate of mixing between the fuel and air rather than the much faster chemical reaction rate.

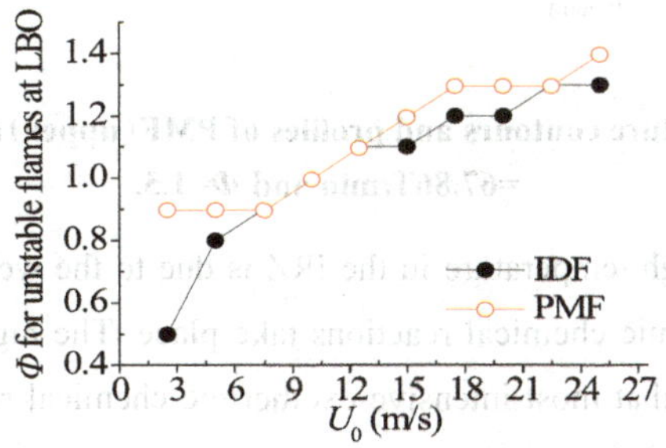

Figure 7.10 Flame stability and lean blow-off limit.

7.3.3 Comparison of flame temperature

The temperature contours of the two swirling flames operating at $\dot{Q}$ =67.86T/min and Φ=1.5 are shown in Figure 7.11. It clearly shows that for both the PMF and IDF, there is a high temperature in the center of the flame, with a variation of less than 200℃ in a large region which extends to around Z=50 mm and it is verified by flow visualization to be the IRZ. The contour designated by the highest temperature is at a certain radial distance from the burner centerline, indicating that the highest temperature is produced in an annular region, consistent with the locus of the bright navy-blue flame front.

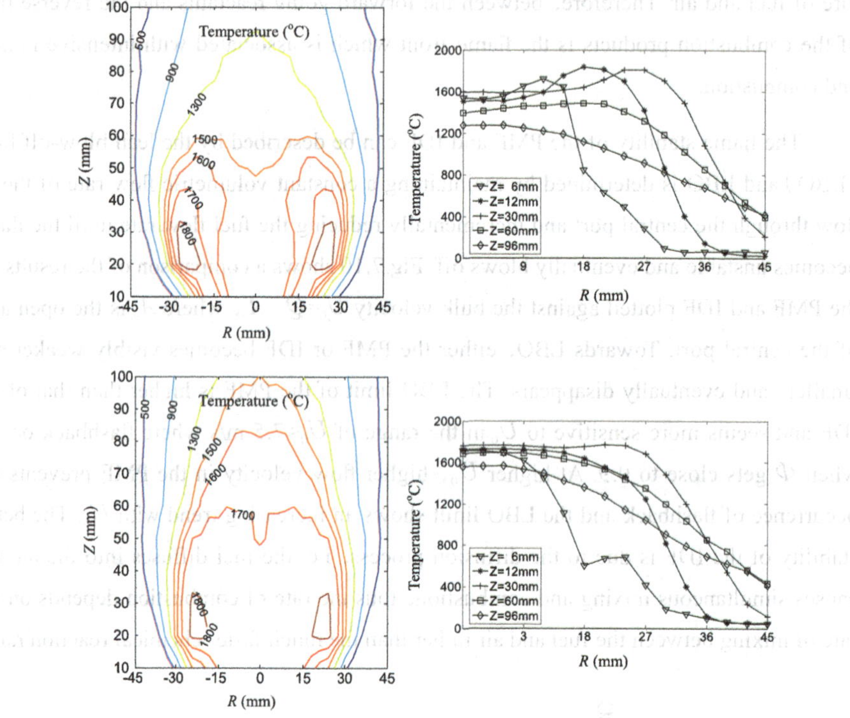

Figure 7.11 Temperature contours and profiles of PMF (upper) and IDF (lower) at $\dot{Q}$ =67.86T/min and Φ=1.5.

The uniform, high temperature in the IRZ is due to the well-stirred reversing flow where intense exothermic chemical reactions take place. The highest temperature at the flame front is evident that most intensive exothermic chemical reactions lies thereunto, i.e. at the edge of the IRZ. This is consistent with Reference (Schmittel *et al.* 2000), which states that the highest temperature is obtained at the boundary of the IRZ and in the IRZ where mainly combustion products prevail, a low heat loss leads to higher temperatures .In the flame boundary outside of the IRZ and especially outside of the flame front, the temperature contours are dense and illustrates that the temperature drops outwards quickly, resulting in steep temperature gradients in Zones 2 and 3.

The temperature profiles of Z=6, 12, 30, 60 and 96 mm are also shown in Figure 7.11. Flow visualization of both the PMF and IDF confirms that the two profiles at Z=6 and 12 mm traverse through Zones 1 and 2. The profile at Z=30 mm penetrates Zones 1 and 3. The other two profiles at Z=60 and 96 mm pass through Zone 3 only. For both the PMF and IDF, the three profiles at Z=6, 12 and 30 mm are smoothly flat and close to each other in the near-centerline region, indicating a uniform temperature distribution in the IRZ. Along each profile, the temperature is the highest at a certain radial distance from the burner centerline, reaffirming that this is where the flame front is. Beyond the flame front and in the flame boundary, the temperature drops outwards radially and shows very steep gradients, also consistent with the temperature contours. Furthermore, from these three profiles which reflect the temperature of the IRZ, it is seen that the IDF achieves a higher temperature (about 200 ℃ higher) than the PMF does, indicating that the IDF which is more diffusional in nature than the PMF generally has a higher flame temperature due to the absence of premixing.

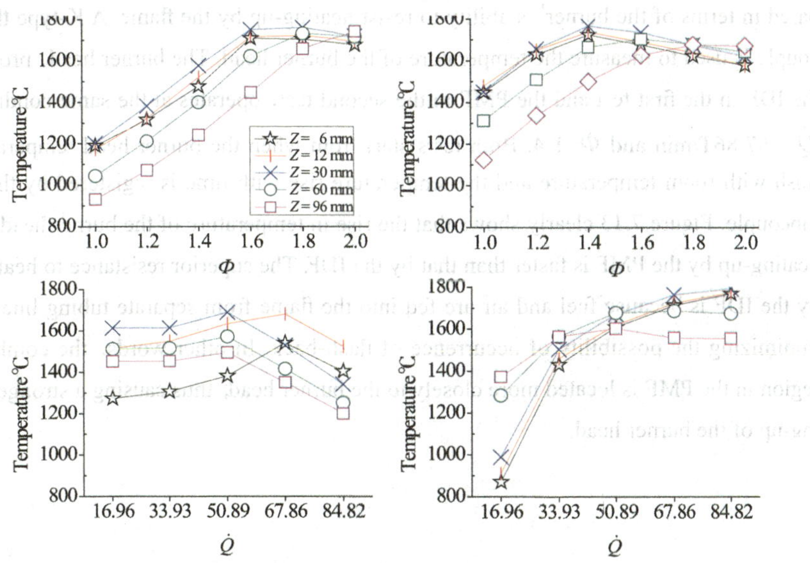

Figure 7.12 Effects of Φ (upper) and $\dot{Q}$ (lower) on centerline temperature of PMF (left) and IDF (right).

Figure 7.12 illustrates the effects of $\dot{Q}$ and Φ on the centerline temperature of the two swirling flames. As previously mentioned, the IRZ is quite large and occupies the majority of the flame by volume. Therefore, the combustion condition in the IRZ can be considered to be representative of the combustion condition of the entire flame. Figure 7.12 shows that at $\dot{Q}$ =67.86T/min, the IRZ attains its peak temperature at Φ=1.6 in the case of the PMF, while at Φ=1.4 in the case of the IDF. At Φ=1.5, the IRZ achieves its peak temperature at $\dot{Q}$ =50.89T/min in the case of the PMF, while at $\dot{Q}$ =84.82T/min in the case of the IDF. The highest flame temperature results from an overall stoichiometric combustion condition. So, it is clear that premixing in the PMF generates a better mixing between fuel and air and leads overall stoichiometric combustion to occur at a higher Φ with fixed $\dot{Q}$ =67.86T/min and a lower with fixed Φ=1.5.

A bad behavior of a burner-flame system is the unavoidable heating-up of the burner itself during the burner's operation. Usually, the burner is heated up by the flame generated by the burner, thus consuming and wasting a portion of the thermal energy produced. A well designed burner-flame system should be capable of effectively transforming energy stored in the fuel into heat and efficiently transferring the heat out of the system for human's useful applications. In this study, the IDF-burner and PMF-burner systems are compared in terms of the burner' s ability to resist heating-up by the flame. A K-type thermocouple is used to measure the temperature of the burner head. The burner head, producing the IDF in the first test and the PMF in the second test, operates at the same condition of $\dot{Q}$ =67.86T/min and Φ=1.4. Both tests start from when the burner head temperature is flush with room temperature and the temperature rise with time is registered by the thermocouple. Figure 7.13 clearly shows that the rise in temperature of the burner head due to heating-up by the PMF is faster than that by the IDF. The superior resistance to heating-up by the IDF is because fuel and air are fed into the flame from separate tubing lines, thus minimizing the possibility of occurrence of flash-back. In other words, the combustion region in the PMF is located more closely to the burner head, thus causing a stronger heating-up of the burner head.

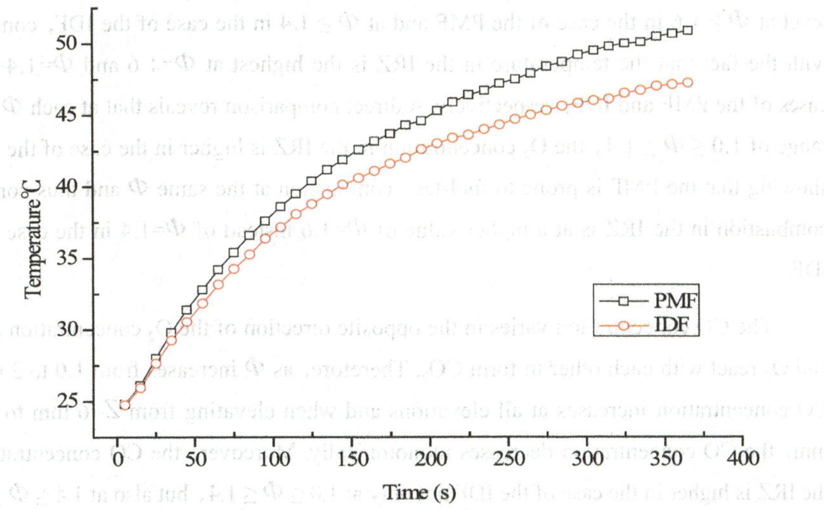

Figure 7.13 Rise in temperature of burner due to heating-up by flames.

7.3.4 Comparison of in-flame gaseous emissions

The composition of the gaseous species including O_2, CO, CO_2, and NO_x in the swirling flames provides information about the combustion process. The dependence of the centerline O_2, CO, CO_2 and NO_x concentrations at Z=6, 12, 30, 60 and 96 mm on Φ are shown in Figure 7.14. The dependence of the centerline O_2, CO, CO_2 and NO_x concentrations at Z=6, 12, 30, 60 and 96 mm on $\dot{Q}$ are shown in Figure 7.15. The effect of Φ is investigated at fixed $\dot{Q}$ =67.86T/min and the effect of $\dot{Q}$ is investigated at fixed Φ=1.5 and both these two conditions correspond to those flame images shown in Figure 7.7 and Figure 7.8, respectively.

It can be observed that the centerline concentration profiles for each species of O_2, CO, CO_2, and NO_x at Z=6, 12 and 30 mm, namely in Zone 1, are close to each other, indicating that the supplied air/fuel, combustion products and entrained ambient air are well mixed in this zone.

Figure 7.14 shows that as Φ is increased from 1.0 to 2.0, the O_2 concentration drops at all elevations due to the increasing fuel supply. When elevating from Z=6 mm to Z=96 mm, the O_2 concentration increases monotonically as more secondary air is entrained into the flame. Because of the well stirred condition in the IRZ, the O_2 concentration profiles at Z=6, 12 and 30 mm are close to each other and much lower than those at Z=60 and 96 mm. This indicates that intensive combustion occurs in the IRZ and more secondary air is entrained into the post-combustion region. The O_2 concentration in the IRZ drops to zero level at $\Phi \geq 1.6$ in the case of the PMF and at $\Phi \geq 1.4$ in the case of the IDF, consistent with the fact that the temperature in the IRZ is the highest at Φ=1.6 and Φ=1.4 in the cases of the PMF and IDF, respectively. A direct comparison reveals that at each Φ in the range of $1.0 \leq \Phi \leq 1.4$, the O_2 concentration in the IRZ is higher in the case of the PMF, showing that the PMF is prone to fuel-lean combustion at the same Φ and thus complete combustion in the IRZ is at a higher value of Φ=1.6 instead of Φ=1.4 in the case of the IDF.

The CO concentration varies in the opposite direction of the O_2 concentration as CO and O_2 react with each other to form CO_2. Therefore, as Φ increases from 1.0 to 2.0, the CO concentration increases at all elevations and when elevating from Z=6 mm to Z=96 mm, the CO concentration decreases monotonically. Moreover, the CO concentration in the IRZ is higher in the case of the IDF not only at $1.0 \leq \Phi \leq 1.4$, but also at $1.4 \leq \Phi \leq 2.0$, indicating a poor CO oxidation process in the case of the IDF.

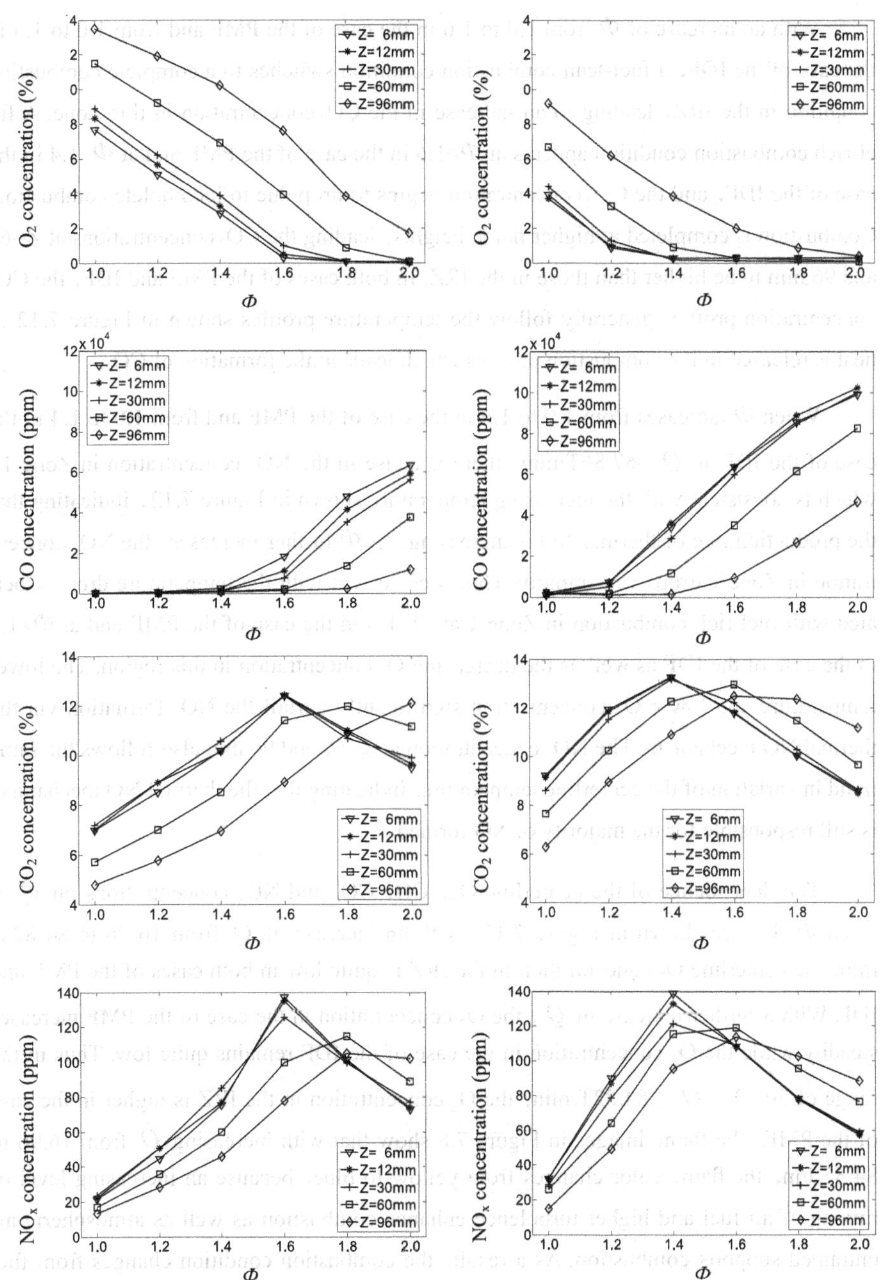

Figure 7.14 Effects of Φ on centerline species concentration of PMF (left) and IDF (right) at $\dot{Q}$=67.86T/min.

With an increase of Φ from 1.0 to 1.6 in the case of the PMF and from 1.0 to 1.4 in the case of the IDF, a fuel-lean combustion condition switches to a complete combustion condition in the IRZ, leading to an increase in the CO_2 concentration in this Zone. A fuel-rich combustion condition appears at Φ>1.6 in the case of the PMF and at Φ>1.4 in the case of the IDF, and the CO_2 concentration begins to drop due to incomplete combustion. Combustion is completed at higher flame heights, leading the CO_2 concentrations at Z=60 and 96 mm to be higher than those in the IRZ. In both cases of the PMF and IDF, the CO_2 concentration profiles generally follow the temperature profiles shown in Figure 7.12 as heat is released in the combustion process which leads to the formation of CO_2.

When Φ increases from 1.0 to 1.6 in the case of the PMF and from 1.0 to 1.4 in the case of the IDF at $\dot{Q}$ =67.86T/min, there is a rise in the NO_x concentration in Zone 1, which is consistent with the increasing temperature shown in Figure 7.12, indicating that the production rate of thermal NO is increasing. As Φ further increases, the NO_x concentration in Zone 1 drops very rapidly. This is consistent with the temperature drop associated with fuel-rich combustion in Zone 1 at Φ>1.6 in the case of the PMF and at Φ>1.4 in the case of the IDF as well as the decreasing O_2 concentration in this region. The lower temperature and lower O_2 concentration significantly curtail the NO_x formation via the thermal NO mechanism. The NO_x concentration at Z=60 and 96 mm also follows the same trend in variation of the centerline temperature, indicating that the thermal NO mechanism is still responsible for the majority of NO formed.

The dependence of the centerline O_2, CO, CO_2 and NO_x concentrations on $\dot{Q}$ at fixed Φ=1.5 are shown in Figure 7.15. With an increase of $\dot{Q}$ from 16.96 to 84.82T/min, the centerline O_2 concentration in the IRZ is quite low in both cases of the PMF and IDF. With a further increase in $\dot{Q}$, the O_2 concentration in the case of the PMF increases steadily while the O_2 concentration in the case of the IDF remains quite low. Thus in the range of $50.89 \leq \dot{Q} \leq 84.82$T/min, the O_2 concentration in the IRZ is higher in the case of the PMF. The flame images in Figure 7.8 show that with increasing $\dot{Q}$ from 16.96 to 84.82 l/m, the flame color changes from yellow to blue, because an increasing level of mixing of air/fuel and higher turbulence enhance combustion as well as atmospheric air entrained supports combustion. As a result, the combustion condition changes from fuel rich to fuel lean. Consequently, the higher O_2 concentration in the range of 50.89T/min $\leq \dot{Q} \leq$ 84.82T/min means that the PMF is more of fuel-lean combustion in nature. The reason is that the premixing present in the PMF augments the level of mixing of the supplied fuel and air.

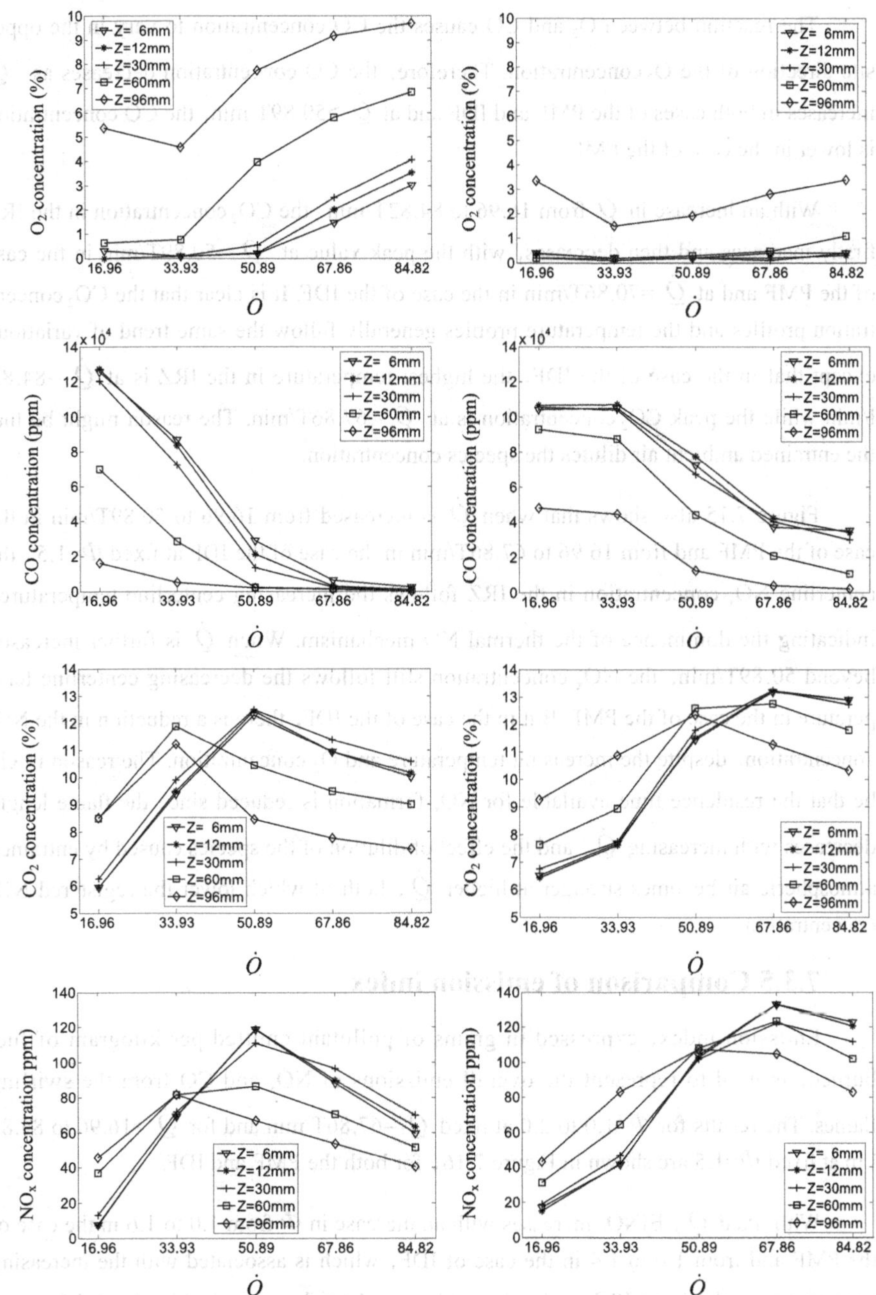

Figure 7.15 Effects of $\dot{Q}$ on centerline species concentration of PMF(left) and IDF(right) at Φ=1.5.

The reaction between O_2 and CO causes the CO concentration to vary in the opposite direction of the O_2 concentration. Therefore, the CO concentration decreases as $\dot{Q}$ increases in both cases of the PMF and IDF and at $\dot{Q}$ >50.89T/min, the CO concentration is lower in the case of the PMF.

With an increase in $\dot{Q}$ from 16.96 to 84.82T/min, the CO_2 concentration in the IRZ firstly increases and then decreases, with the peak value at $\dot{Q}$ =50.89T/min in the case of the PMF and at $\dot{Q}$ =70.86T/min in the case of the IDF. It is clear that the CO_2 concentration profiles and the temperature profiles generally follow the same trend of variation, except that in the case of the IDF, the highest temperature in the IRZ is at $\dot{Q}$ =84.82 l/min while the peak CO_2 concentration is at $\dot{Q}$ =67.86T/min. The reason might be that the entrained ambient air dilutes the species concentration.

Figure 7.15 also shows that when $\dot{Q}$ is increased from 16.96 to 50.89T/min in the case of the PMF and from 16.96 to 67.86T/min in the case of the IDF at fixed Φ=1.5, the centerline NO_x concentration in the IRZ follows the increasing centerline temperature, indicating the dominance of the thermal NO mechanism. When $\dot{Q}$ is further increased beyond 50.89T/min, the NO_x concentration still follows the decreasing centerline temperature in the case of the PMF. But in the case of the IDF, there is a reduction in the NO_x concentration, despite the increasing temperature and O_2 concentration. The reason might be that the residence time available for NO_x formation is reduced since the flame length decreases with increasing $\dot{Q}$, and the effect of dilution of the species caused by entrained atmospheric air becomes stronger at higher $\dot{Q}$, both of which lower the registered NO_x concentration.

7.3.5 Comparison of emission index

Emission index, expressed in grams of pollutant emitted per kilogram of fuel burned, is used to represent the overall emissions of NO_x and CO from the swirling flames. The results for Φ=1.0 to 2.0 at fixed $\dot{Q}$ =67.86T/min and for $\dot{Q}$ =16.96 to 84.82 l/m at fixed Φ=1.5 are shown in Figure 7.16, for both the PMF and IDF.

With fixed $\dot{Q}$, $EINO_x$ increases with an increase in Φ from 1.0 to 1.6 in the case of the PMF and from 1.0 to 1.4 in the case of IDF, which is associated with the increasing temperature in the large IRZ as the flame changes from lean-combustion to stoichiometric-combustion in the IRZ, indicating that the thermal NO mechanism controls the overall NO_x emission, due to the presence of the large-size, high-temperature IRZ in the two swirling flames. As Φ increases beyond 1.6 in the case of the PMF and beyond 1.4 in the case of the IDF, $EINO_x$ does not follow the decreasing temperature in the IRZ but keeps nearly constant. The flame length increases with increasing Φ, thus the residence time available for NO_x formation becomes longer, producing more NO_x and making $EINO_x$ at a constant level.

With fixed $\dot{Q}$, EICO drops quickly from Φ=1.0 to 1.4 in both cases of the PMF and IDF, and changes very little at higher Φ. At Φ=1.0, the flame length of either the PMF or IDF is very short with the post combustion region being negligibly small. CO formed in the flame cannot be effectively oxidized in such a small Zone 3, resulting in a high value of EICO. As Φ increases to 1.2, a double flame structure, which consists of a lower flame involving Zone 1 and Zone 2 and an upper diffusion flame including Zone 3, becomes more discernible and meanwhile Zone 3 becomes longer. Thus, the oxidation of CO by the upper diffusion flame is fast and hence a lower EICO is formed. Further drops in EICO occur until the complete combustion condition is established in Zone 1 at Φ=1.6 in the case of the PMF. When Φ further increases, the fuel-rich combustion condition sets up in Zone 1 and the higher CO production rate in Zone 1 is probably balanced by the higher oxidation rate in Zone 3, leading to a nearly constant EICO.

With fixed Φ=1.5 and increasing $\dot{Q}$ from 16.96 to 84.82T/min, $EINO_x$ in both cases of the PMF and IDF exactly follows the variation of the temperature in the IRZ, which shows that the thermal NO mechanism is controlling the overall NO_x emission.

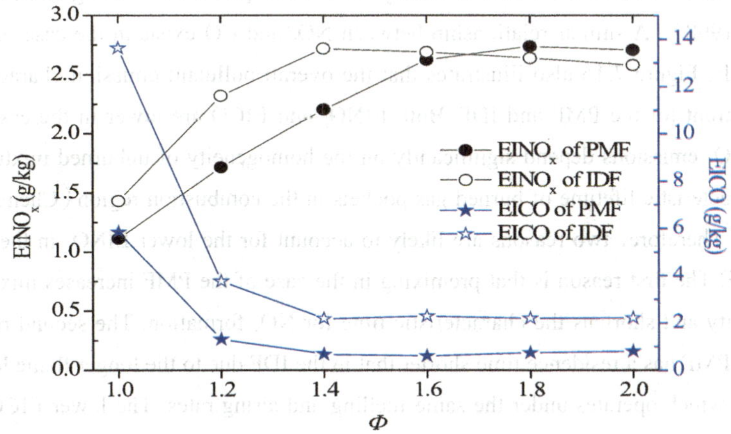

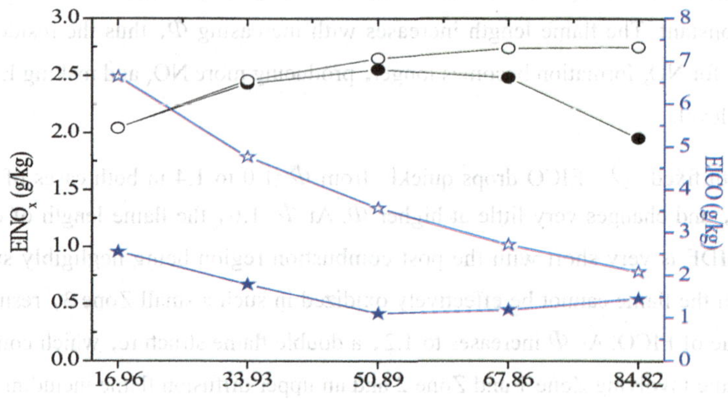

Figure 7.16 Emission indices of PMF and IDF.

With fixed Φ=1.5 and increasing $\dot{Q}$ from 16.96 to 84.82T/min, EICO in both cases of the PMF and IDF decreases monotonically. The reason is that the higher jet velocity and higher turbulence associated with higher $\dot{Q}$ enhance combustion, leading to a reduction in the amount of CO formed in the flame. Moreover, the increasing turbulence enhances entrainment of ambient air to help further oxidation of CO in the post-combustion region.

In the present study of the IDF, a reduction in NO_x is generally accompanied by an increase in CO emissions, which is usually seen in non-premixed swirling flames (Cheng *et al.*, 1998). A similar relationship between NO_x and CO exists in the case of IDF in this study. Figure 7.16 also illustrates that the overall pollutant emission characteristics are different for the PMF and IDF. Both EINO$_x$ and EICO are lower in the case of the PMF. NO_x emissions depend significantly on the homogeneity of unburned mixtures and the characteristic lifetime of burned gas pockets in the combustion region (Cheng *et al.*, 1998). Therefore, two reasons are likely to account for the lower EINO$_x$ in the case of the PMF. The first reason is that premixing in the case of the PMF increases mixture homogeneity and shortens the characteristic time for NO_x formation. The second reason is that the PMF has a residence time shorter that in the IDF due to the longer flame length of the IDF which operates under the same fuelling and airing rates. The lower EICO in the case of the PMF is ascribed to two reasons. One is that premixing in the case of the PMF increases the oxidation rate of CO. The other is that the relatively high O_2 concentration supports the conversion of CO to CO_2.

CHAPTER 8 NUMERICAL SIMULATION OF THE ISOTHERMAL FLOW

Numerical simulation is now truly on par with experiment and theory as a research tool to produce multi-scale information that is not available by using any other technique. In this study, the jet flow of the swirling IDF in the non-reactive conditions is simulated. There are two purposes for carrying out the numerical simulation. Firstly, the experimental investigations covered by Chapters 3to7 provide little information on the flow field and only a limited number of parameters can be measured. The numerical approach helps in drawing a detailed picture on the fluid-mechanics of the jet flow that cannot be measured in this study. Secondly, it helps to account for the internal structure of the flame and a comparison between the computational and experimental results can be made to address for the differences and show the validity of the numerical approach.

The numerical simulation of the swirling IDF is a problem of turbulent flame modeling. Veynante and Vervisch (2002) mentioned that the modeling of a turbulent flame involves a wide range of coupled problems. On the one hand, fluid properties of the combustion system must be known to describe the mixing between reactants and all the transfer phenomena occurring in the flame. On the other hand, detailed chemical reaction schemes are necessary to estimate the consumption rate of the fuel, the formation of combustion products and pollutant species. So a precise knowledge of the chemistry is required to predict ignition, stabilization and extinction of reaction zones together with pollution. The problem becomes more complex if two or three phase systems are encountered or radiative heat transfer is taken into account. However, since our objective is to examine the fluid-dynamical characteristics of the flame, hence the numerical simulation can be simplified to only model the non-reactive fluid flow. The focus of the analysis of the computational results is to address three issues related to the isothermal fluid flow, i.e. velocity field, air/fuel mixing and swirl strength.

8.1 Background

Due to the devoid of flow field measurement, i.e. the experimental derivation of flow velocity either by intrusive measurements or by the non-intrusive diagnostic techniques such as Laser-Doppler anemometry (LDA), Laser-Doppler Velocimetry (LDV) and Particle Imaging Velocimetry (PIV), little information about the flow velocity field is acquired in the experimental investigation part of this project study. Therefore, it is necessary to perform numerical simulations of the swirling IDF to provide desired information about velocity, air/fuel mixing and swirl strength.

Thanks to the rapid development of super-computers and numerical methods, the application of Computational Fluid Dynamics (CFD) has become a popular tool for predicting flow phenomena and for designing complex 3-D flow systems in which swirling flows are indispensable. The solution of flow problems which require real gas modeling also has become feasible for simple flow configurations and meantime numerous research activities have been devoted to the numerical simulation of combustion and particularly to flame modeling. However, the ability to computationally predict turbulent swirling flames remains elusive. Additionally, there are important unresolved issues relevant to turbulent swirling flames, i.e. the effects of the interaction between the flow and chemistry and the emission of pollutants. Compared with simple jet flames, pilot-stabilized flames and bluff-body-stabilized flames, the validation of numerical tools for swirling flames, form the next level of difficulty because intense swirl promotes the formation of complex, recirculating flow structures and leads to a strong coupling between the flow and the chemical kinetics (Masri *et al.*2000).

For simplification purpose, only the isothermal conditions are considered in this project study to provide a basic picture of the flow field of the swirling IDF. Hence, the governing equations for the investigated problem are the conservation equations of mass and momentum for an incompressible Newtonian fluid in an open swirling jet. When considering an incompressible flow, the equation for conservation of mass is:

$$\nabla \cdot V = 0 \tag{8.1}$$

where V is the fluid velocity and ∇ is the del operator.

For a Newtonian and incompressible fluid flow, the equation for conservation of momentum is the so-called Navier-Stokes equation:

$$\rho(\frac{\partial V}{\partial t} + V \cdot \nabla \cdot V) = -\nabla p + \mu \nabla^2 V \tag{8.2}$$

ρ is the fluid density, p is the pressure and μ is the dynamic viscosity. Note that the body force term is omitted in Equation 8.2 since the effect of gravity on gas jet flow is negligible.

A solution of the Navier-Stokes equation is a velocity field or flow field, which is a description of the velocity of the fluid at a given point in space and time. Once the velocity field is solved for, other quantities of interest may be found. The solution of the governing Equations 8.1 and 8.2 does not raise any fundamental difficulties in the case of inviscid or laminar flows. However, the simulation of turbulent flows presents a significant problem. One feasible approach is to approximate the governing equations by the Reynolds-Averaged Navier-Stokes equations (RANS), and use turbulence models which model the Reynolds stresses to close the RANS equations. Next, the principles of solution of the governing equations will be further elaborated on. (Blazek, 2001)

An overwhelming majority of the numerical methods for the solution of the RANS equations employ a separate discretisation in space and in time. First of all, the physical space, where the flow is to be computed is divided into a large number of geometric elements. There are three main categories of spatial discretisation methods, finite difference, finite volume and finite element, among which, the finite volume method is nowadays very popular and in wide use. Whichever spatial discretisation scheme we might select, it is important to ensure that the scheme converges to the solution of the discretised equations. In a further step, the resulting time-dependent equations are advanced in time, starting from a known initial solution, with the aid of a suitable method. The temporal discretisation schemes are divided into explicit schemes and implicit schemes. The former is numerically cheap but its employment is restricted because of stability limitations. The latter has the advantages of superior robustness and convergence speed in the case of stiff simulations, turbulence modeling or in the case of highly stretched grids. In addition, the implicit schemes have no problem of stability. It is earlier mentioned that turbulence models must be supplemented to close the RANS equations for simulating turbulent flows. The turbulence models can be classified into first-and second-order closures, among which first-order closures are more widely used. The first-order closures can be further categorized into zero-, one-, and multiple-equation models, among which only the one – and two-equation models take history effects into account. The most popular one-equation model is the Spalart and Allmaras model and among the large number of two-equation models, the K-ε model and the K-ω model are most often used. Regardless of the numerical methodology chosen to solve the governing equations, we have to specify suitable initial and boundary conditions. The initial conditions determine the state of the fluid at the time t=0, or at the first step of an iterative scheme. The initial solution should satisfy at least the governing equations. Any computational domain is only a certain "truncated" part of the physical domain and the truncation creates artificial boundaries where values of the physical quantities have to be specified. It is important that such boundary conditions be constructed properly so that the solution on the truncated domain stays as close as possible to the solution which would be obtained for the whole physical domain (Harvard and Thomas 1999).

The software named FLUENT is a powerful and flexible general-purpose CFD code used for engineering simulations of all levels of complexity. It offers a comprehensive range of physical models that can be applied to a broad range of industries and applications. The package is based on the finite volume method on a collocated grid and has a geometric modeling and grid generation tool called GAMBIT. In this project study, the software FLUENT and GAMBIT are utilized to perform the numerical simulation of the flow field of the swirling IDF in non-reacting conditions. The software TECPLOT is used to post-process the computational results.

8.2 Mesh generation

Figure 8.1 displays the geometric model of the swirl burner.

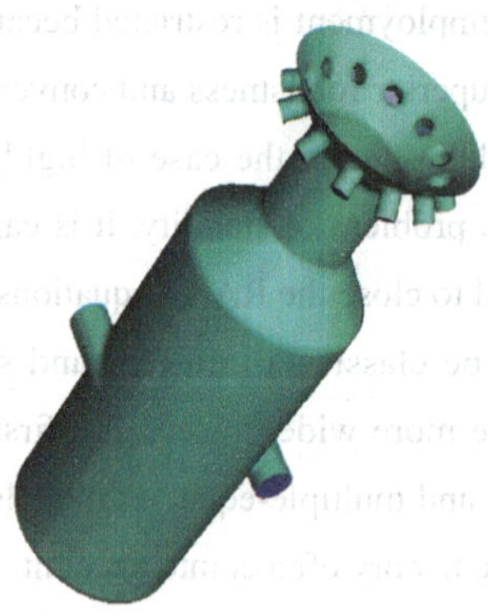

Figure 8.1 Geometric model of the swirl burner.

The swirling IDF consists of a central swirling air jet and twelve fuel jets that are circumferentially arranged around the air jet. Since the objective is to simulate these jet flows in cold conditions without combustion, the fuel jets are replaced by air jets for the purpose of simplification. Due to the geometric complexity of the swirl burner, mixed unstructured mesh becomes necessary to offer flexibility in the treatment of complex geometries. The meshing of this model is illustrated in Figure 8.2.

Figure 8.2 Mesh of the swirl burner.

Figure 8.2 shows that a mixed tetrahedral and hexahedral grid cells are placed on the model. Hexahedral grid cells are used in the regular geometries to alleviate the demand on the computational power and tetrahedral grid cells are chosen to accurately resolve the flow where the geometries join each other or evolve irregular changes.

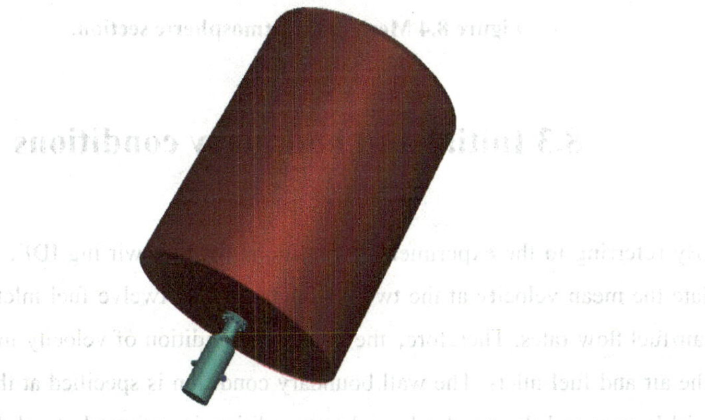

Figure 8.3 Geometric model of the entire computational domain.

Besides the mesh of the swirl burner, the computational domain also comprises a large cylindrical section starting from the burner exit, representing the physical domain of atmosphere in which the jet flow or combustion develops. The geometric model and meshing of this section are shown in Figure 8.3 and Figure 8.4, respectively. As a result of its large size, only hexahedral grid cells are used to shorten the time required for simulation and the computational results show that this mesh is not too coarse to increase the numerical errors significantly. The number of grid is 545161.

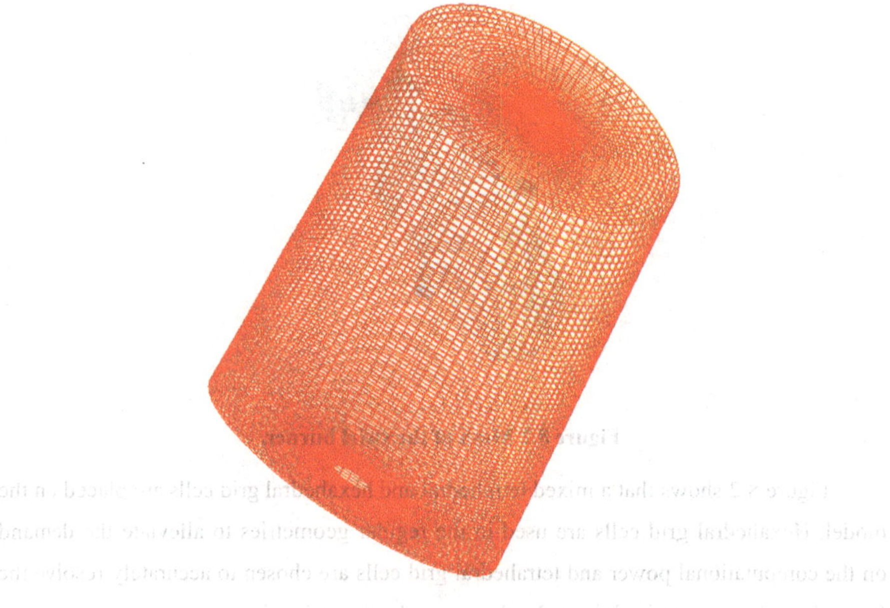

Figure 8.4 Mesh of the atmospheric section.

8.3 Initial and boundary conditions

By referring to the experimental conditions for the swirling IDF, we can directly calculate the mean velocity at the two air inlets and the twelve fuel inlets from the supplied air/fuel flow rates. Therefore, the boundary condition of velocity inlet is defined as both the air and fuel inlets. The wall boundary condition is specified at the solid walls of the swirl burner and the no-slip boundary condition is enforced at solid regions by default. The boundary condition of pressure outlet is placed at the atmospheric section. For specification of the initial condition, the axial component of flow velocity in the whole computational domain is set to 1 m/s, and it is found that this initial condition accelerates convergence efficiently.

8.4 Turbulence models

When modeling turbulent flows with a significant amount of swirl, the advanced turbulence models of the RNG K-ε model, the realizable K-ε model and the Reynolds stress model are usually recommended for use (Fluent User Guide). In order to make an appropriate choice of the turbulence model, several computations are run by trial and error, and compare the results from using the standard K-ε model, the RNG K-ε model and the realizable K-ε model. The comparison indicates that only the realizable K-ε model yields satisfactory results, with the other two models resulting in either divergence or meaningless flow velocities. The computation is first carried out considering the steady conditions, and then unsteady conditions are taken into account because the swirling jet flow is associated with strong oscillation called precessing vortex core (PVC).

8.5 Computational results

8.5.1 Velocity vector

The non-reacting swirling IDF at the condition of Re=8000 and Φ=2.0 is simulated. Shown in Figure 8.5 is the 2-D flow field of velocity vectors in the near region downstream from the burner's divergent nozzle. It is seen that a large IRZ is established inside the swirling jet flow. The IRZ or the reverse flow is identified by negative vectors and it is clear that there are two vortex centers which reside symmetrically to the burner centerlinc and just above the burner exit. The central region between the vortex centers has intense negative vectors which indicate almost uniform axial downward velocity profile and the reverse flow extends into the air port. In the narrow boundary region of the swirling jet flow which has a conical shape, there are high positive vectors. The reverse flow in the IRZ propagates against the out-flowing expanding jet flow, producing an axis-symmetrical stagnation surface whose shape is visible in Figure 8.5 as a white line where the velocity vectors are vanishing. The location of the forward stagnation point is at a distance of around one diameter of the air port inside the air port. At larger distances from the burner nozzle, the stagnation surface extends outward radially along the inner side of the IRZ. Under reacting conditions, the flame front should be positioned near the stagnation surface, where reactants and products will mix. The lengths and arrows of vectors in Figure

8.5 indicate the magnitude and direction of velocity, respectively.

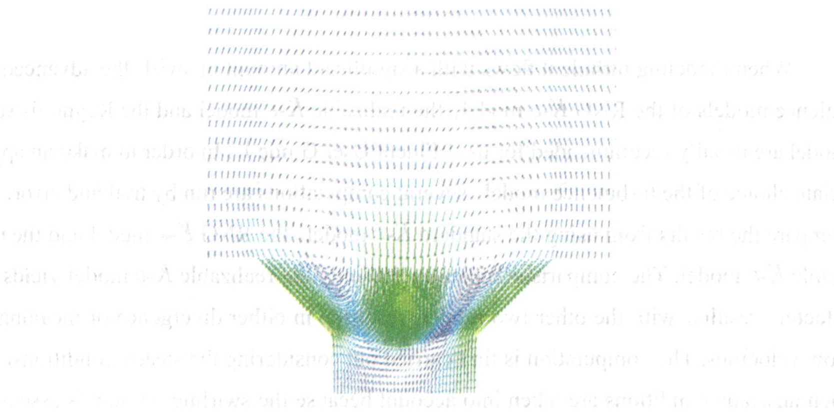

Figure 8.5 Near field downstream of the burner nozzle.

8.5.2 Air/fuel mixing

The vertical plane showing the near field in Figure 8.5 does not reveal any fuel inlet. In order to investigate the physics of air/fuel mixing, another vertical plane involving two fuel inlets is shown in Figure 8.6. It is observed that no penetration of the fuel jets into the IRZ occurs, due to the small fuel-air momentum ratio. The simulation performed at the condition of Re=8000 and Φ=1.0 yields the same finding, simply because in the case of Φ=1.0, the fuel-air momentum ratio is even smaller. On the other hand, there are inward vectors on the outskirts of the swirling jet flow, indicating that ambient air is entrained into the jet flow.

Figure 8.6 Another near field downstream of the burner nozzle.

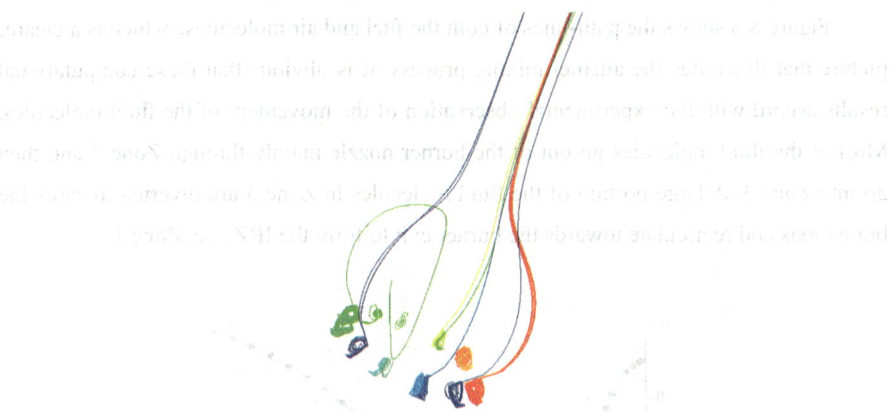

Figure 8.7 Path-lines of the fuel molecules.

To further examine the air/fuel mixing, it is necessary to check the trajectories of the fuel molecules, which are revealed by the path-lines shown in Figure 8.7. It is observed that the fuel molecules bounce back when impinging onto the swirling air jet because the momentum of the fuel jet is very small in comparison with that of the air jet, thus a recirculation zone forms in the zone inside the fuel port and near to the fuel port exit. After quite a few cycles of recirculation, the fuel molecules would diffuse into the air and hence the process of air/fuel mixing occurs. The mixing process of reactants and recirculated products occurs along the conical stagnation surface. From the theoretical perspective, the processes of air/fuel mixing and combustion take place simultaneously, and the flame front is formed in the shear layer between the air and fuel flow where air/fuel mixing occurs.

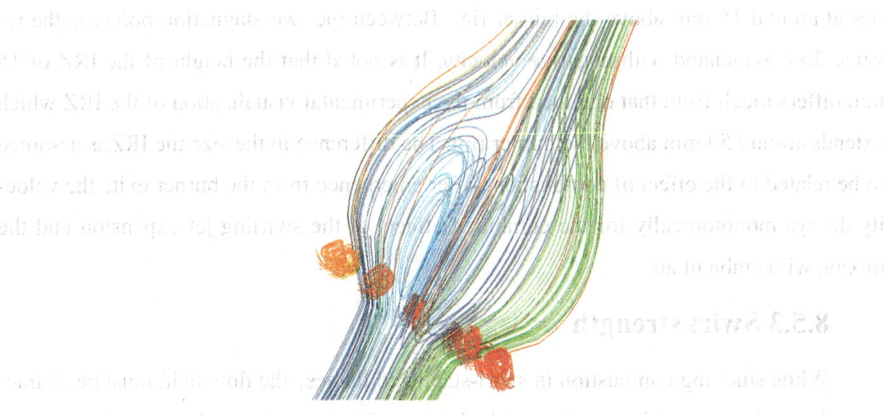

Figure 8.8 Path-lines of the fuel and air molecules.

Figure 8.8 shows the path-lines of both the fuel and air molecules, which is a clearer picture that illustrates the air/fuel mixing process. It is obvious that these computational results accord with the experimental observation of the movement of the fluid molecules. Most of the fluid molecules go out of the burner nozzle mainly through Zone 2 and then go into Zone 3. A large portion of the fluid molecules in Zone 3 are diverted towards the burner axis and recirculate towards the burner exit to form the IRZ, i.e. Zone 1.

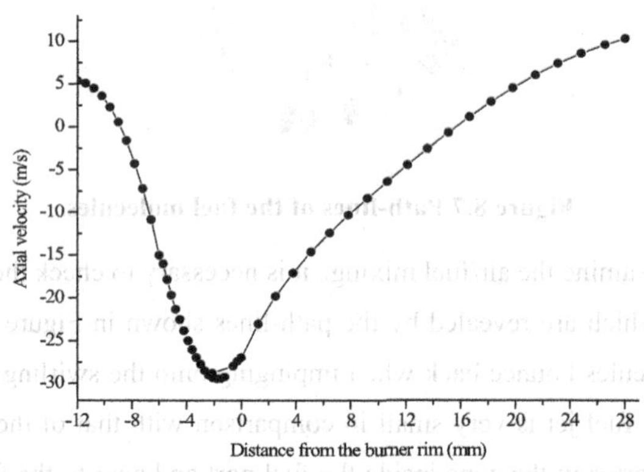

Figure 8.9 Axial velocity profile along the burner axis.

To better understand the mixing between reactants and products, the velocity distribution along the burner axis is investigated and the result is shown in Figure 8.9. Two stagnation points where the velocity is zero are found. The forward stagnation point is located at an axial distance of around 9 mm below the burner rim and the backward one lies at around 15 mm above the burner rim. Between the two stagnation points is the reverse flow associated with negative velocity. It is noted that the height of the IRZ of 15 mm differs much from that obtained from the experimental visualization of the IRZ which extends around 50 mm above the burner rim. The difference in the size the IRZ is assumed to be related to the effect of combustion. After emergence from the burner exit, the velocity decays monotonically for the combined effects of the swirling jet expansion and the mixing with ambient air.

8.5.3 Swirl strength

While studying combustion in swirl-stabilized flames, the flow field must be characterized and also the swirl strength must be known. Here, we rely on the numerical simulation to get an idea of the degree of swirl. The definition of swirl number is given in Equation 2.2, which puts that the swirl number can be calculated from the axial and tangential velocity profiles. Practically, there is a decay in the generated swirl as the swirling jet flow moves downstream in the atmosphere. Therefore, the swirl number is usually evaluated just above the burner exit. Figure 8.10 shows the velocity vector plot at the horizontal plane of Z=0 mm where the swirl strength is evaluated, namely at the cross-sectional plane of the rim of the burner' s divergent outlet.

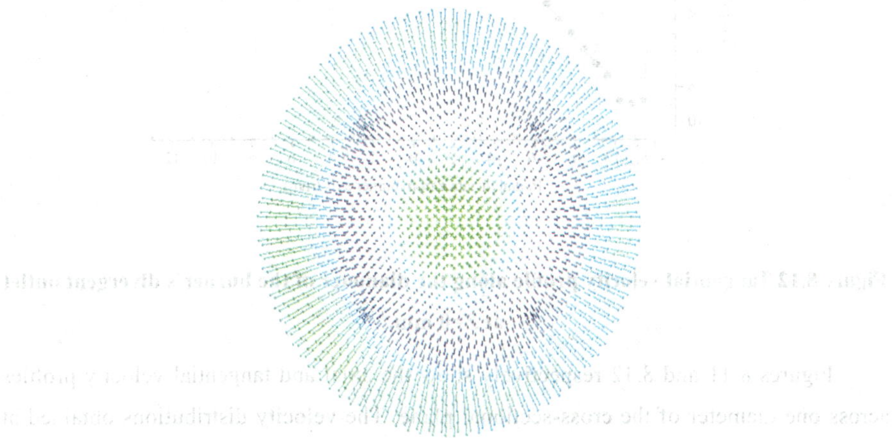

Figure 8.10 Axial velocity profile along the diameter of the burner's divergent outlet at *Z*=0 mm.

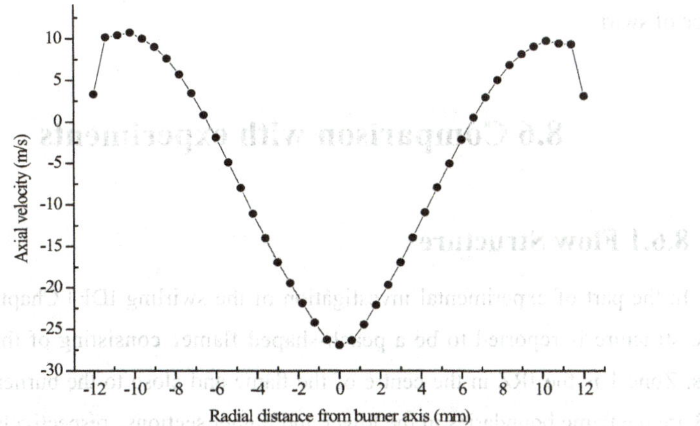

Figure 8.11 Axial velocity profile along the diameter of the burner's divergent outlet at *Z*=0 mm.

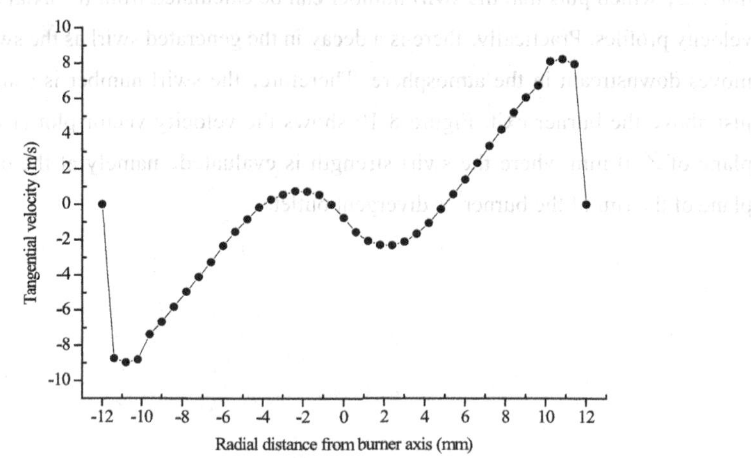

Figure 8.12 Tangential velocity profile along the diameter of the burner's divergent outlet at Z=0 mm.

Figures 8.11 and 8.12 respectively show the axial and tangential velocity profiles across one diameter of the cross-sectional plane. The velocity distributions obtained at other diameters are found to be similar. Based on Equations 2.2, 2.3 and 2.4, the swirl number is calculated to be 0.66. The simulation conducted at Re=8000 and Φ=1.0 yields a larger value of S=0.74. The reason probably is that when the fuel jets with smaller momentum impinge onto the air jet, less energy loss is produced and hence results in a higher degree of swirl.

8.6 Comparison with experiments

8.6.1 Flow Structure

In the part of experimental investigation of the swirling IDF (Chapters 3to7), the flame structure is reported to be a peach-shaped flame, consisting of three distinctive zones. Zone 1 is the IRZ in the centre of the flame and close to the burner exit. Zones 2 and 3 are the flame boundaries in the lower and higher sections, respectively. By using the flow visualization technique, the location of vortex eyes and the motion of the fluid molecules inside the flow field are also reported. Namely, fluid molecules coming out of the burner mainly effuse through Zone 2 into Zone 3. Then a large portion of the fluid molecules in Zone 3 are diverted towards the burner axis and recirculate towards the burner exit to form Zone 1. All these experimental observations can be summarized in the schematic shown in Figure 8.13.

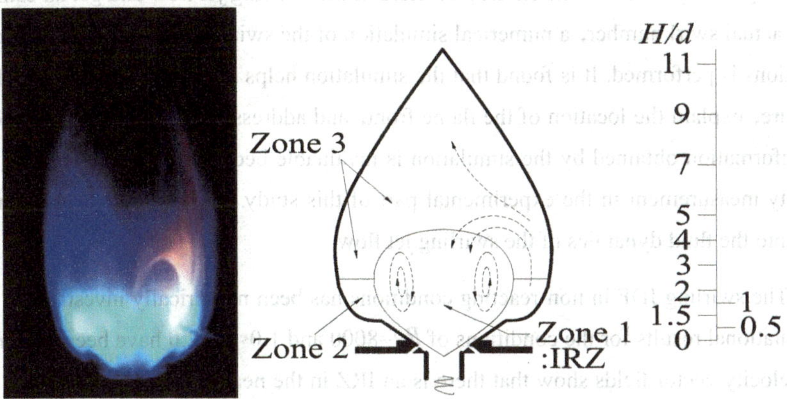

Figure 8.13 Experimental flow structures.

It can be witnessed that in general the flow field in Figure 8.13 is similar to the computational one in Figure 8.5. One difference is that the size of the IRZ is larger in the experimental case where the vortex eyes are located around $2d$ above the burner exit and the centre-to-centre spacing in between is around $2d$. While in the computational case, the vortex eyes are much closer to the burner exit and the two vortex centers are also much closer, such that the size of the IRZ is smaller compared with the experimental one. The reason for this difference might be due to the influence of combustion on the IRZ. It is assumed that the presence of combustion enlarges the recirculation zone significantly.

8.6.2 Swirl Number

For the experimental part of this study, the swirl strength is estimated based on a geometric swirl number. No actual swirl number is given due to the difficulty in velocity measurements. However, we know from the literature review that for strongly swirling flames which are associated with flow recirculation, the actual swirl number generally exceeds 0.6. Therefore, the comparison of swirl strength can be made between the literature and the numerical simulation. The actual swirl number yielded by the computation is 0.74, which is larger than 0.6 and verified that the numerical approach used is reasonable.

8.7 Summary of numerical simulation

To get more information on the flow field in the swirling jet flow and get an estimate of the actual swirl number, a numerical simulation of the swirling IDF under non-reacting conditions is performed. It is found that the simulation helps account for the 3-zone flame structure, explain the location of the flame front, and address the air/fuel mixing process. The information obtained by the simulation is invaluable because there has been no flow velocity measurement in the experimental part of this study. The simulation gives an insight into the fluid dynamics of the swirling jet flow.

The swirling IDF in non-reacting conditions has been numerically investigated. The computational results for the conditions of Re=8000 and $1.0<\Phi<2.0$ have been presented. The velocity vector fields show that there is an IRZ in the near flow field, consistent with the experimental observation. The direction of velocity vectors is also in good accordance with the observed movement of fluid molecules in the three-zone flame structure. The examination of the trajectories of fuel molecules reveals that no penetration of the fuel jets into the IRZ occurs, thus air/fuel mixing is by the diffusion of fuel into the air jet. Additionally, the numerical simulation yields a proper estimation of swirl number. It concludes that the numerical approach used is capable of generating reasonable velocity data and hence many other flow properties which are not experimentally measured can be derived from the velocity data. The disadvantage of the numerical approach is that it is hard on computer resources and thus more accurate simulation demands more computational time. Furthermore, the simulation of the fluid flow inside the swirl chamber produces quite large dissimilarity.

CHAPTER 9 CONCLUSIONS AND RECOMMENDATION

9.1 Conclusions

The thermal, emission and heat transfer characteristics of a swirling inverse diffusion flame (IDF) burning liquefied petroleum gas (LPG) have been studied experimentally. The swirling IDF operates in a high swirl stabilization mode and is associated with the formation of an internal recirculation zone (IRZ). In comparison with the IDF without induced swirl, the flame length is shortened and the flame stability is enhanced by the presence of the IRZ. The identified governing parameters of the swirling IDF are air jet Reynolds number *Re*, overall equivalence ratio Φ, geometric swirl number S' and nozzle-to-plate distance *H*. Experimental investigations have been conducted on the characteristics of the swirling IDF including flame appearance, flame structure, flame temperature, in-flame stable gaseous species, overall pollutant emissions and flame impingement heat transfer. Comparisons of the swirling IDF with other flames have also been performed. The main conclusions from this study are drawn in the following.

9.1.1 Burner testing

A swirl burner exploiting two tangential entries of air to induce high degrees of swirl has been designed, fabricated and optimized. The geometric configuration of the burner has been chosen for the purpose of minimization of the total pressure drop, including the considerations of tangential entry of fluid, position of the tangential nozzles, and addition of a throat in "Y" shape and a divergent outlet. In the light of the formation of a sufficiently large IRZ, the considerations involve the divergent outlet, the large-diameter central air port and the short divergent exit.

Using LPG as the fuel, the burner produces a highly stable swirling IDF. Within the range of $2000<Re<10000$ and the range of $1.0<\Phi<2.0$, detailed experimental studies have been performed regarding flame appearance, flame structure, flame temperature, in-flame gaseous species concentration, overall pollutant emissions and flame impingement heat transfer.

9.1.2 Flame appearance and structure

When the swirling IDF is formed at Re <4000, the flame consists of two parts: a peach-shaped flame root and a long flame tail. The flame root encloses an IRZ and the IRZ is only observable at Re >1000. At Re >6000, the long flame tail totally disappears and the peach-shaped flame root remains. Therefore, the flame root can characterize the flame structure of turbulent swirling IDFs.

The flame structure consists of three distinctive zones: Zones 1, 2 and 3. Zone 1 is the IRZ, i.e. a reverse flow in the centre of the flame and close to the burner exit. Zone 2 is the flame boundary in the lower section, being in contact with ambient air on the outer side and with Zone 1 on the inner side. Zone 2 is always navy-blue in color, indicative of premixed combustion and Zone 2 is the mixing zone in which the swirling air jet coming out from the air port initially mixes with the fuel coming out from the fuel jets. Zone 3 is the flame boundary in the upper section and overlaps Zones 1 and 2. Zone 3 acts as a source of reversed fluid molecules in the IRZ.

9.1.3 Flame temperature

The size of the IRZ is quite large and the temperature in Zone 1 is high and uniform because of the well-stirred situation in the IRZ where intense exothermic chemical reactions take place. However, the highest flame temperature occurs at the lower edge of the IRZ, i.e. the interface between Zones 1 and 2 which is confirmed to be the flame front by its color of bright navy-blue. The large-size Zone 1 makes itself capable of being representative of the entire flame in terms of the combustion condition.

The effects of the governing parameters of Re and Φ on the centerline temperature show that as Φ increases from 1.0 to 2.0 at Re=8000, the combustion condition in Zone 1 or in the whole flame changes from fuel-lean to fuel-rich with the stoichiometric combustion occurring at Φ=1.4.

9.1.4 In-flame stable gaseous species

The composition of the gas within the swirling IDF is mainly composed of polyatomic stable gaseous species of O_2, CO_2, CO and NO_x and each of them has a uniform concentration distribution in Zone 1 and the concentration distribution is coupled well with the combustion condition. Generally, the CO concentration changes to the contrary of the variation of O_2 concentration, due to the reaction between O_2 and CO to form CO_2. The increase or decrease in the CO_2 concentration follows the flame temperature because heat is mainly released in the combustion process which results in the formation of CO_2. As for NO/NO_x emissions, the formation of NO increases with temperature and the NO concentration decreases very rapidly when the flame temperature is below 1800 K. This reveals that the thermal NO mechanism is controlling the NO/NO_x emissions and the role played by the prompt NO mechanism is not significant.

9.1.5 Overall pollutant emissions

The emission indices of NO_x and CO are considered because they are the main totally emitted pollutants which are hazardous to human body. At the range of $2000<Re<10000$ and the range of $1.0<\Phi<2.0$, $EINO_x$ varies from 1.43 to 2.75 k/kg. EICO varies from 2.08 to 13.64 k/kg. The data reveal that the swirling IDF achieves a moderate level of NO_x emission and an ultra low level of CO emission under certain operational conditions.

9.1.6 Heat transfer

The heat transfer characteristics of the swirling IDF as an impinging flame are investigated by a heat flux sensor. The swirling effect is found to influence heat transfer in three ways.1. The heat transfer at the stagnation point is deteriorated by swirl.2. Both high impinging velocity and high flame temperature dwell in the flame boundary, and the local heat flux is the highest at a certain radial position, following the position of either the peak impinging velocity or the peak flame temperature.3. As H increases, the swirling jet brings the position of h_{max} gradually farther away from the stagnation point.

Under the condition of Re=8000 and Φ=1.4, there is an optimum nozzle-to-plate distance for the highest overall heat transfer rate. With $\dot{q}$ =4.27 kW at $H_{optimum}/d$=1.5, the heat transfer efficiency is 62.5%. The effect of Φ on the heat transfer is that an increase in Φ leads to higher overall heat transfer rates, but the heat transfer efficiency increases and then decreases with further increasing Φ. The effect of Re on the heat transfer is that an increase in energy input leads to higher overall heat transfer rates and the heat transfer efficiency decreases with increasing Re. The effect of S' on the heat transfer is that an increase in S' leads to lower overall heat transfer rates and lower heat transfer efficiencies, thus showing an unfavorable effect of swirl on heat transfer.

9.1.7 Comparison of flames

9.1.7.1 Swirling and non-swirling IDFs

The comparison of swirling IDFs and non-swirling IDFs shows that due to the presence of the IRZ, the swirling IDFs are larger in cross-sectional diameter and shorter in flame length. Additionally, the flame stability of swirling IDFs is much better than IDFs without swirl and non-swirling IDFs are more diffusional in nature because more intensive mixing and hence more complete combustion occurs in IDFs with swirl.

Mostly, $EINO_x$ of non-swirling IDFs is lower than that of swirling IDFs. One reason is that a higher temperature is expected in swirling IDFs due to better combustion and the presence of the large-size high-temperature IRZ. The other reason is that swirling IDFs might have a residence time of an equivalent level to that for non-swirling IDFs, based on the assumption that under the same conditions, the decrease in residence time caused by shorter flame length in swirling IDFs might be offset by the increase in residence time that is caused by the recirculating flow.

EICO of non-swirling IDFs is higher than that of swirling IDFs at $\Phi \geq 1.2$ and Re=8000, which means that the combustion is better in swirling IDFs due to the stronger air/fuel mixing in the IRZ. In the range of 2000< Re <10000 at Φ=1.5, EICO is higher in the case of non-swirling IDFs, which once again indicates that the combustion is more intense and more complete in swirling IDFs.

For non-swirling IDFs and at small H, there exists a cool core in the centre of the flame. The cool core mainly consists of air supplied from the central air port and leads to very low heat flux in the stagnation region. When swirl is introduced, swirl promotes rapid mixing of the air/fuel and boosts intense combustion so that the cool core present in the case of non-swirling IDFs disappears. Thus, swirling IDFs can achieve complete combustion at lower H, compared with non-swirling IDFs. The comparison of the highest overall heat transfer rates of swirling and non-swirling IDFs at their individual $H_{optimum}$ shows that the swirl has an adverse effect on heat transfer.

9.1.7.2 Swirling IDF and swirling PMF

Two different types of swirl-stabilized flames have similar flame shape and size. Both PMF and IDF are peach-shaped flames attached to the divergent nozzle of the burner and enclose a large strong recirculation zone which is located along the burner axis. The flame structure of the two flames are also the same because both Re and S', which are key parameters representative of the fluid-dynamic characteristics of the swirling jet flow effusing from the central port, do not undergo significant changes between the cases of PMF and IDF. IDF is more diffusional in nature due to the absence of premixing between fuel and air before combustion and thus the flame length of IDF is slightly larger than that of PMF.

Both IDF and PMF are assumed to be stabilized by the IRZ. However, PMF is less stable when compared with IDF. The reason might be related to the intricate transient chemical reactions or the difference in chemical species and radicals in PMF and IDF.

The data of flame temperatures of IDF and PMF indicate that premixing in PMF generates a better mixing of fuel and air and leads totally stoichiometric combustion to occur at a higher Φ with fixed $\dot{Q}$ =67.86 l/m and a lower $\dot{Q}$ with fixed Φ=1.5.

The comparison of $EINO_x$ and EICO of IDF and PMF indicates that both $EINO_x$ and EICO are lower in PMF because premixing in PMF increases mixture homogeneity and shortens the characteristic time for NO_x formation, and also because PMF has a residence time shorter than that in IDF due to the longer flame length of IDF. The lower EICO in PMF is ascribed to two reasons. One is that premixing in PMF increases the oxidation rate of CO. The other is that the relatively high O_2 concentration supports the conversion of CO to CO_2.

9.1.7.3 Potential application of swirling IDF

It is apparent that swirling IDFs have many advantages over non-swirling IDFs. Compared with non-swirling IDFs, the compact combustion in swirling IDFs yields a shorter and wider flame, which does not vary in dimension as wildly as a non-swirling IDF with either changing fuel supply or air supply. Further, swirling IDFs are more stable and thus have wider range of operation, being outstandingly superior to the narrow operational range for non-swirling IDFs. These advantages associated with swirling IDFs yield a potential application of swirling IDFs in industrial furnaces, for instance, to generate useful heat by burning waste material. The features of high flame temperature and ultra-low CO emission further favor the application of swirling IDFs in furnaces.

The experimental finding shows that when utilizing the post-combustion zone of a swirling IDF to heat a target plate, the radial heat flux distribution is rather uniform. Therefore, another potential application of swirling IDFs is for uniform heating.

9.2 Recommendation for future work

(1) While studying combustion in swirl-stabilized flames, the flow field and the swirl strength are always of interest. It is necessary to measure the velocity profiles and static pressure profiles across the swirling jet flow. Some techniques available include

LDV for measuring velocity profiles and PIV for measuring the 2-D flow field. Once we are able to get these data, we can characterize the flow field in both non-reacting (isothermal or cold) conditions and reacting conditions and thus study the influence of combustion on the flow field. For example, the reversed mass flow rate can be obtained by cross section integration of measured axial velocity and temperature (density) profiles. On the other hand, from the data of velocity pressure profiles, the actual swirl number can be evaluated at the inlet of the quarl because the geometric swirl number is quite different from the actual swirl number due to the decay in the generated swirl as the flow moves downstream.

(2) Swirling flames are currently investigated by many researchers to resolve the flow structure and stability characteristics. In the present study, the flow visualization technique for estimating the size and shape of the IRZ is too rough and the comparison of PMF and IDF points out that the reason for the difference in flame stability is still unknown. Hydroxyl radical is usually used as a flame marker in non-premixed and partially premixed flames. Therefore, our future work may use the LIF (laser induced fluorescence) technique to measure reactive species concentrations such as OH. On the one hand, based on the data of [OH], we can accurately locate the positions of the flame front and the IRZ. The effects of the governing parameters on the size and shape of the IRZ thus can be figured out. On the other hand, the difference in stabilization mechanisms betneen PMF and IDF which might be related to the intricate transient chemical reactions can be further understood.

(3) Many researchers have successfully operated combustion with ultra-low NO_x emission. To lower $EINO_x$ of the present swirling IDF, many control strategirg should be explored and tested, such as using excess air for fuel-lean combustion, change the degree of swirl for lower flame temperatures, vary the fuel injection typology for strong and rapid mixing of fuel and air, lower inlet air temperature, etc.

REFERENCES

[1]]Barr, J. Diffusion flames. Fourth Symposium (International) on Combustion, 1953: 765-771.

[2] Barr, J. Length of Cylindrical laminar diffusion flames. Fuel, 1954: 33, 35-41.

[3] Becker, H.A. and Yamazaki, S. Entrainment, momentum flux and temperature in vertical free turbulent diffusion flames. Combustion and Flame, 1978: 33, 123-149.

[4] Bindar, Y. and Irawan A. Size and structure of LPG and hydrogen inverse diffusion flames at high level of fuel excess. Sixth Asia-Pacific International Symposium on Combustion and Energy Utilization, 2002: 124-130.

[5] Blazek, J. Computational Fluid Dynamics: Principles and Applications. Alstom Power Ltd., Baden-Daettwil, Switzerland, 2001.

[6] Blevins, L.G., Mulholland, G.W. and Davis, R.W. Carbon monoxide and soot formation in inverse diffusion flames. Fifth International Microgravity Combustion Workshop, Cleveland, 1999.

[7] British standard. Measurement of Fluid Flow in Closed Conduits. BS 1042: Section 2.1, 1983.

[8] Brohez, S., Delvosalle, C. and Marlair, G. A two-thermocouples probe for radiation corrections of measured temperatures in compartment fires. Fire Safety Journal, 2004: 39, 399-411

[9] Burke, S.P. and Schumann, T.E. Diffusion flames. Industrial & Engineering Chemistry, 1928: 20, 988-1004.

[10] Butcher, S.S., Rao, U., Smith, K.R., Osborn, J., Azuma, P. and Fields, H. Emission factors and efficiencies for small-scale open biomass combustion: toward standard measurement techniques. The Annual Meeting of the American Chemical Society, 1984.

[11] Cha, M.S., Lee, D.S. and Chung, S.H. Effect of swirl on lifted flame characteristics in nonpremixed jets. Combustion and Flame, 1999: 117, 636-645.

[12] Chan, C.K., Lau, K.S., Chin, W.K. and Cheng, R.K. Freely propagating open premixed turbulent flames stabilized by swirl. Twenty-Fourth Symposium (International) on Combustion, The Combustion Institute, 1992: 511-518.

[13] Cheng, R.K., Yegian, D.T., Miyasato, M.M., Samuelsen, G.S., Benson, C.E., Pellizzari, R. and Loftus, P. Scaling and development of flow-swirl burners for low-emission furnaces and boilers. Proceedings of the Combustion Institute, 2000: 28, 1305-1313.

[14] Cheng, T.S., Chao, Y.C., Wu, D.C., Yuan, T., Lu, C.C., Cheng, C.K. and Chang, J.M. Effects of fuel-air mixing on flame structures and NO_x emissions in swirling methane jet flames. Twenty-Seventh Symposium (International) on Combustion, The Combustion Institute, 1998: 1229-1237.

[15] Chou, C.P., Chen, J.Y., Yam, C.G. and Marx, K.D. Numerical modeling of NO formation in laminar Bunsen flames-A flamelet approach. Combustion and Flame, 1998: 114, 420-435.

[16] Claypole, T.C. and Syred, N. The effect of swirl burner aerodynamics on NO_x formation. Proceedings of the eighteenth International Symposium on Combustion, The Combustion Institute, 1981: 81-89.

[17] Dong, L.L., Cheung, C.S. and Leung, C.W. Heat transfer characteristics of an impinging inverse diffusion flame jet – Part I: Free flame structure. Inter-

national Journal of Heat and Mass Transfer, 2007: 50, 5108-5123.

[18] Dong, L.L., Cheung, C.S. and Leung, C.W. Heat transfer characteristics of an impinging inverse diffusion flame jet. Part II: Impinging flame structure and impingement heat transfer. International Journal of Heat and Mass Transfer, 2007: 50, 5124-5138.

[19] Echigo, R., Nishiwaki, N. and Hirata, M. A study on the radiation of luminous flames. Eleventh Symposium (International) on Combustion, 1967: 381.

[20] Feikema, D., Chen, R.H. and Driscoll, J.F. Blowout of nonpremixed flames: Maximum coaxial air velocities achievable, with and without swirl. Combustion and Flame, 1991: 86, 347-358.

[21] Feikema, D., Chen, R.H. and Driscoll, J.F. Enhancement of flame blowout limits by the use of swirl. Combustion and Flame, 1990: 80, 183-195.

[22] Fleck, B.A. Experimental and Numerical Investigation of the Novel Low Nitrogen Oxide CGRI Burner. PhD Thesis, the Queen's University, 1998.

[23] Friend, J.N. The Chemistry of Combustion. London: Gurney & Jackson, 1992.

[24] Gupta, A.K., Lilley, D.G. and Syred, N. Swirl Flow. Abacus Press, Tunbridge Wells, England, 1984.

[25] Harvard, L. and Thomas H.P. Fundamentals of Computational Fluid Dynamics. NASA Ames Research Center, 1999.

[26] Holman, J.P., Heat transfer. New York: McGraw Hill, c2010.

[27] Hottel, H.C. and Hawthorne, W.R. Flame and explosions. Third Symposium (International) on Combustion, 1949: 254-266.

[28] Huang, L. and El-Genk, M.S. Heat transfer and flow visualization experiments of swirling, multi-channel, and conventional impinging jets. International Journal of Heat and Mass Transfer, 1997: 41, 583-600.

[29] Huang, R.F., Yang, J.T. and Lee, P.C. Flame and flow characteristics of double concentric jets. Combustion and Flame, 1997: 108, 9-23.

[30] Huang, X.Q., Leung, C.W., Chan, C.K. and Probert, S.D. Thermal characteristics of a premixed impinging circular laminar-flame jet with induced swirl. Applied Energy, 2006: 83, 401-411.

[31] Hwang, S.S. and Gore, J.P. Characteristics of combustion and radiation heat transfer of an oxygen-enhanced flame burner. Proceedings of the Institution of Mechanical Engineers, Journal of Power and Energy, 2002: 379-386.

[32] Ji, J. and Gore, J.P. Flow structure in lean premixed swirling combustion. Proceedings of the Combustion Institute, 2002: 861-867.

[33] Kaplan, C.R. and Kailasanath, K. Flow-field effects on soot formation in normal and inverse methane-air diffusion flames. Combustion and Flame, 2001: 124, 275-294.

[34] Kwok, L.C., Leung, C.W. and Cheung, C.S. Heat transfer characteristics of an array of impinging pre-mixed slot flame jets. International Journal of Heat and Mass Transfer, 2005: 48, 1727-1738.

[35] Lee, D.H., Won, S.Y., Kim, Y.T. and Chung, Y.S. Turbulent heat transfer from a flat surface to a swirling round impinging jet. International Journal of Heat and Mass Transfer, 2002: 45, 223-227.

[36] Lee, G.Y., Jurng, J. and Hwang, J. Synthesis of carbon nanotubes on a catalytic metal substrate by using an ethylene inverse diffusion flame. Letters to the Editor / Carbon, 2004: 42, 682-685.

[37] Lockwood, F.C. and Moneb H.A. Fluctuating temperature measurements in turbulent jet diffusion flame. Combustion and Flame, 1982: 47, 291-314.

[38] Makel, D.B. and Kennedy, I.M. Soot formation in laminar inverse diffusion flame. Combustion Science and Technology, 1993: 81, 207.

[39] Masri, A.R. Pope, S.B. and Dally, B.B. Probability density function computations of a strongly swirling nonpremixed flame stabilized on a new burner. Proceedings of the Combustion Institute, 2000: 28, 123-132.

[40] Mikofski, M.A., Williams, T.C., Shaddix, C.R. and Blevins, L.G. Flame height measurement of laminar inverse diffusion flames. Combustion and Flame, 2006: 14663-72.

[41] Mitchell, R.E. Experimental and numerical investigation of confined laminar diffusion flames. Combustion and Flame, 1980: 37, 227-244.

[42] Ng, T.K., Leung, C.W. and Cheung, C.S. Experimental investigation on the heat transfer of an impinging inverse diffusion flame. International Journal of Heat and Mass Transfer, 2007: 50, 3366-3375.

[43] Olivani, A., Solero, G., Cozzi, F. and Coghe, A. Near field flow structure of isothermal swirling flows and reacting non-premixed swirling flames. Experimental Thermal and Fluid Science, 2007: 31, 427-436.

[44] Owsenek, B.L., Cziesla, T., Mitra, N.K. and Biswas, G. Numerical investigation of heat transfer in impinging axial and radial jets with superimposed swirl. International Journal of Heat and Mass Transfer, 1997: 40, 141-147.

[45] Peters, N. Laminar flamelet concept in turbulent combustion. Twenty-First Symposium (International) on Combustion, The Combustion Institute, 1986: 1232-1250.

[46] Ropper, F.G. The prediction of laminar jet diffusion flame size: Part I, Theoretical model. Combustion and Flame, 1977: 29, 219-226.

[47] Savage, L.D. The enclosed laminar diffusion flame. Combustion and Flame, 1962: 6, 77-87.

[48] Schmittel, P., Gunther, B., Lenze, B., Leuckel, W. and Bockhorn, H. Turbulent swirling flames: experimental investigations of the flow field and formation of nitrogen oxide. Proceedings of the Combustion Institute, 2000:

28, 303-309.

[49] Sidebotham, G..W. and Glassman, I. Flame temperature, fuel structure, and fuel concentration effects on soot formation in inverse diffusion flames. Combustion and Flame, 1992: 90, 269-283.

[50] Smith, S.R. and Gordon, A.S. Studies of diffusion flames. I. The methane diffusion flame. The Journal of Physical Chemistry, 1956: 60, 759-763.

[51] Sobiesiak, A. and Wentzell, J.C. Characteristics and structure of inverse flames of natural gas. Proceedings of the Combustion Institute, 2002: 30, 743-749.

[52] Sunderland, P.B., Krishman, S.S. and Gore, J.P. Effects of oxygen enhancement and gravity on normal and inverse laminar jet diffusion flames. Combustion and Flame, 2004: 136, 244-256.

[53] Syred, N. and Beer, J.M. Combustion in swirling flows: a review. Combustion and Flame, 1974: 23, 143-201.

[54] Syred, N., Chigier, N.A. and Beer, J.M. Flame stabilization in recirculation zones of jets with swirl, Proceedings of the seventeenth International Symposium on Combustion, The Combustion Institute, 1971: 563.

[55] Sze, L.K., Cheung, C.S. and Leung, C.W. Appearance, temperature and NO_x emission of two inverse diffusion flames with different port design. Combustion and Flame, 2006: 144, 237-248.

[56] Sze, L.K., Cheung, C.S. and Leung, C.W. Temperature distribution and heat transfer characteristics of an inverse diffusion flame with circumferentially arranged fuel ports. International Journal of Heat and Mass Transfer, 2004: 47, 3119-3129.

[57] Takagi, T., Xu, Z. and Komiyama, M. Preferential diffusion effects on the temperature in usual and inverse diffusion flames. Combustion and Flame, 1996: 106, 252-260.

[58] Tangirala, V., Chen, R.H. and Driscoll, J.F. Effect of heat release and swirl on the recirculation within swirl-stabilized flames. Combustion Science and Technology, 1987: 51, 75 – 95.

[59] Terasaki, T. The effects of fuel-air mixing on NO_x formation in non-premixed swirl burners. Twenty-Sixth Symposium (International) on Combustion, The Combustion Institute, 1996: 2733-2739.

[60] Tremeer, G.B. and Jawurek, H.H. The "hood method" of measuring emissions of rural cooking devices. Biomass Bioenergy, 1999: 16, 341–345.

[61] Veynante, D. and Vervisch, L. Turbulent combustion modeling. Progress in Energy and Combustion Science, 2002: 28, 193-266.

[62] Ward, J. and Mahmood, M. Heat transfer from a turbulent, swirling, impinging jet. Proceedings of the seventh International Heat Transfer Conference, 1982: 401-407.

[63] Ward, J. and Mahmood, M. The effect of swirl on mass/heat transfer from arrays of turbulent, impinging jets. Proceedings of ASME-WAM, 1993: 57-64.

[64] Wentzell J.C. The Characteristics and Structure of Inverse Flames of Natural Gas. MSc Thesis, the Queen's University , 1998.

[65] William, P., Partridge, J. and Laurendeau, N.M. Nitric oxide formation by inverse diffusion flames in staged-air burners. Fuel, 1995: 74, 1424-1430.

[66] William, P., Partridge, J., Reisel, J.R. and Laurendeau, N.M. Laser-saturated fluorescence measurements of nitric oxide in an inverse diffusion flame. Combustion and Flame, 1999: 116, 282-290.

[67] Wolfhard, H.G. and Parker, W.G. A spectroscopic investigation into the structure of diffusion flames. Proceedings of the Physical Society, 1945: 1949-1957.

[68] Wu K.T. and Essenhigh, R.H. Mapping and structure of IDF of methane. Twentieth Symposium (International) on Combustion, The Combustion Institute, 1984: 1925-1932.

[69] Wu, K.T. The Comparative Structure of Normal and Inverse Diffusion Flames. PhD Dissertation, the Ohio State University, 1984.

[70] Yagi, S. and Iino, H. Radiation form soot particles in luminous flames. Combustion and Flame, 1962: 6, 77-87.

[71] Yuan, Z.X., Chen, Y.Y., Jiang, J.G. and Ma, C.F. Swirling effect of jet impingement on heat transfer from a flat surface to CO_2 stream. Experimental Thermal and Fluid Science, 2006: 31, 55-60.

[72] Zhao, Q.W., Chan, C.K. and Zhao, H.F. Numerical simulation of open swirl-stabilized premixed combustion. Fuel 83, 2004: 1615–1623.

[73] Zhao, Z., Yuen, D.W., Leung, C.W. and Wong, T.T. Thermal performance of a premixed impinging circular flame jet array with induced-swirl. Applied Thermal Engineering, 2008: 29, 159-166.